Who Killed the Great Auk?

Great Auks courting by Jan Wilczur

Who Killed the Great Auk?

JEREMY GASKELL

UNIVERSITY PRESS

OXFORD
UNIVERSITY PRESS

Oxford University Press, Great Clarendon Street, Oxford OX2 6DP

Oxford New York

Athens Auckland Bangkok Bogotá Bombay Beunos Aires
Calcutta Cape Town Dar es Salaam Delhi Florence Hong Kong
Istanbul Karachi Kuala Lumpur Madras Madrid Melbourne
Mexico City Nairobi Paris Singapore Taipei Tokyo Toronto

and associated companies in
Berlin Ibadan

Oxford is a trade mark of Oxford University Press

Published in the United States
by Oxford University Press Inc., New York

A catalogue record for this book is available from the British Library

Library of Congress Cataloging in Publication Data
(Data available)

Gaskell, Jeremy.
Who killed the great Auk?/Jeremy Gaskell; with original illustrations by Jan Wilczur
Includes bibliographical references
1. Great auk. 2. Extinct birds. I. Title

QL696.C42 G34 2000 333.95′833–dc21 00–044623

ISBN 0 19 856478 3

1 3 5 7 9 10 8 6 4 2

Typeset by EXPO Holdings, Malaysia

Printed in Great Britain on acid-free paper by
Biddles Ltd, *www.biddles.co.uk*

Acknowledgements

It is difficult to thank adequately all those who have contributed in various ways to this work, but I must make special mention of the following: Dr Arni Einarsson of the Institute of Biology, Reykjavik and Sjófu Kristjánsdóttir of the Manuscript Department of the National Library of Iceland, both of whom gave valuable advice on the translation of an important historical document; Hallgrímur Gunnarsson and his wife Kristbjörg for their kindness and hospitality and for organizing excursions during my visit to Iceland; their son Gunnar Hallgrímsson who helped with the finer points of Icelandic grammar and orthography and with a number of important research enquiries; John Smith of the University of Westminster and Valentina Arena for advice on seventeenth-century Latin; Mike Mullen and Frank Lange with whom I sat up into the small hours translating from German; Dr Bill Bourne of Aberdeen University who made comments on the manuscript and with whom I enjoyed a lively correspondence subsequently; Richard Clark who solved my computer problems; and Juan Herrera who furnished me with a serviceable hard disk after one near disaster.

Those who have given of their time to help with particular enquiries are Roger Norris of the Dean and Chapter Library, Durham; Lindy Brewster of the Durham University Archaeology Museum; Clive Powell and Brian Thynne of the National Maritime Museum, Greenwich; Les Jessop of the Hancock Museum, Newcastle-upon-Tyne; Alwyne Wheeler, former editor of the *Archives of Natural History*; Jørgen Nielsen and Professor Jon Fjeldså of the Zoologisk Museum, University of Copenhagen; Vidar Bakken of the Zoological Museum, University of Oslo; and Joan Ritcey of the Centre for Newfoundland Studies, Memorial University, St John's. I should also like to express heartfelt thanks to members of staff at the British Library who were ever willing to go the extra mile in tracking down primary source material. Every effort has been made to trace the original copyright holders. Any omissions notified to Oxford University Press will be corrected in subsequent printings. Kate Kilpatrick has proved an ideal and sagacious commissioning editor, deftly guiding the book through the various stages through which any publishing proposal must pass.

For kind permission to use original photographs I extend my thanks to Professor David Nettleship, Gunnar Hallgrímsson, and Reinhard Ruge and, for supplying other illustrative material, to Dr Linda Birch of the Edward Grey Institute, Oxford. Jan Wilczur kindly agreed at short notice to furnish his masterly frontispiece. Errol Fuller not only allowed me unlimited access to all the illustrations which he had collated over four years for his own magnificent work on the Great Auk but also designed the dust jacket from one of his own evocative paintings. There can be few instances of generosity towards a relatively unknown author to rival Errol's and the debt I owe him is truly incalculable.

Researching a book of this kind is no light undertaking so I would, finally, express my warm appreciation to my friends Alenka Lawrence, Joanna Bogle, and Colonel John Deverell for their advice and encouragement at critical stages in the preparation of the manuscript.

Contents

Illustrations

Introduction

Anyone who happened to be looking out to sea from one of the remote Scottish islands on a spring day in the early nineteenth century might have been fortunate enough to catch sight of 'the King (or Queen) of the Auks' as it dived for fish just offshore at high water. The Great Auk was one of our largest sea fowl — as large in fact as the Great Northern Diver to which it bore some resemblance. However, a clear view of the bird side on would have revealed the massive bill and seemingly incongruous white spot on the otherwise dark head — both features which distinguished it readily enough from any other sea bird of comparable size or colour.

The Great Auk has the unhappy distinction of being the only bird that was once a regular visitor to the coasts of the British Isles but is no longer to be seen. However, it would be a mistake to think that there was anything inevitable about its disappearance. When the species became extinct in the mid-nineteenth century, the first laws for the universal protection of wildfowl were already under consideration on both sides of the Atlantic, and if just a few of this once abundant bird had survived for a few decades more, subsequent generations could have looked forward to seeing Great Auks in both northern European waters and along the eastern seaboard of North America.

Over the centuries so many creatures have suffered progressive decimation of their numbers through incessant slaughter that we are apt to forget that, as a rule, it is difficult to drive animals and birds to extinction by that means alone: in the great majority of cases the loss or degradation of natural habitat is the primary cause of a species' disappearance from the face of the earth. To see the truth of this, it is only necessary to consider those birds of prey in Europe which a century ago were persecuted throughout much of their range but which, during the four years of respite granted by the First World War, achieved a remarkable recovery. In contrast to most birds of prey, the Great Auk was confined on account of its flightlessness to a strictly limited number of breeding places, a factor which renders any such species vulnerable to over-exploitation. However, the rocky islets on which it once struggled ashore each year to lay and brood its single egg were so remote from centres of civilization that, even today, they can be approached only under the most favourable weather conditions. We can understand how, in these circumstances, a large flightless bird might have suffered a reduction, perhaps severe, in its numbers, if persistently exploited as a food source. But fowling expeditions for this purpose alone could hardly have led to complete extinction, for rough seas or thick fog would sometimes have prevented such expeditions, thereby allowing the birds some successful breeding seasons.

In his celebrated nineteenth-century children's story, *The Water Babies*, Charles Kingsley gave a sympathetic portrait of the 'Gairfowl' (as the Auk was known locally in northern Europe), in terms which suggest that he too thought that the ultimate fate of the species was difficult to account for:

> 'Once we were a great nation and spread over all the northern isles: but men shot us so, knocked us on the head and took our eggs till at last there were none of us left except off the old Gairfowlskerry, just off the Iceland coast. Even there we had no peace — for one day the land rocked, and the sea boiled, and the sky grew dark, and all the air was filled with dust, and down tumbled old Gairfowlskerry into the sea. Some of us were dashed to pieces and some drowned, but those who were left got away to Eldey; but the dovekies tell me they are all dead now, and that another Gairfowlskerry has arisen out of the sea close to the old one, but that

Fig. 1 The Great Auk on the all-alone stone. Illustration by Linley Sambourne for the 1884 edition of *The Water Babies* by Charles Kingsley.

> it is such a poor flat place that it is not safe to live on and so here am I left all alone.' This was the gairfowl's story, and, strange as it may seem, it is every word of it true.

Some of the nineteenth-century authors, apparently seeking to absolve mankind of complete responsibility for the disappearance of the species, emphasized such natural calamities as the submarine volcanic eruption described here by Charles Kingsley. Periodic eruptions were, however, natural hazards for the Great Auk, as all the islands on which it bred were, or had been at one time, volcanic. The effect of the last eruption in the Auk's history, as Charles Kingsley implies, was merely to drive the surviving birds nearer to human habitations so that the species' complete extinction came about ten or twelve years earlier than might otherwise have been the case.

The last standard bird identification guide in which the Great Auk was figured and described was the first volume of James Fisher's *Bird Recognition*, which appeared in 1947. Since that time, understandably enough, both authors and publishers have felt that the inclusion of the Great Auk in modern field guides would be misplaced. However, the move to relegate the Great Auk to a series of learned (and some not so learned) monographs dating back to the middle of the nineteenth century commenced with the publication of T.A. Coward's *Birds of the British Isles* in 1920. He wrote:

> From time to time...popular interest is aroused in the Great Auk: an egg changes hands and brings a high price in a London saleroom, and a more or less incorrect account of the Garefowl appears in the Press. It is the money value of the egg...and not the bird that creates interest.

Coward was an ornithologist — a true bird*watcher* — who delighted in birds as living, vibrant life-forms. To him the desire of the collector simply to possess seemed, to say the least, anachronistic. Paradoxically, this distaste led Coward to give the reader short shrift and to refer anyone interested to 'the extensive literature of the Great Auk.' However, this literature suffers considerably from having been written too close to the period with which it is most concerned — the decades immediately prior to the Great Auk's extinction in European waters. What we are left with is a series of incomplete accounts of the species, many of which gave contemporaries the misleading impression that the *locus classicus* of the Great Auk was in the Old World, rather than in the New.

There is a second, more serious charge to be levelled at most of the nineteenth-century authors which stems from the fact that news of the Great Auk's demise coincided with the dissemination of the theory of evolution cogently argued by Charles Darwin. The phrase 'the survival of the fittest' was soon on everyone's lips, but Darwin's contemporaries (and indeed others in later years), frequently misunderstood what this naturalist was actually saying. Too often 'the survival of the fittest' was construed, not in the sense of the 'survival of the best suited', but as no more than a polite form of the piratical excuse 'might is right'. Indeed, 'adapt or die' seemed to become the uncomfortable gospel of the age and it is interesting to observe the ways in which Christian clergymen endeavoured to harmonize it with their own. We may take as typical the views of the Revd Moses Harvey of Newfoundland, as expressed in the May 1874 issue of the New York publication *Forest and Stream*, a journal devoted to both natural history and sporting interests:

> It is evident that in the 'battle of life', such a bird as the Great Auk had but a poor chance. In a word, where competition for available provisions is so keen, where the 'struggle for existence' is so terrible, where only the 'fittest' survive, such a simpleton as the Great Auk must ere long be gobbled up. When the fat 'innocent at home' actually walked into the mouths of its foes — great gawk that it was — its doom must be annihilation sooner or later. Such proved to be the case.

Ill-digested Darwinism such as this came to be regarded as a rule of life about which people were thought to do well not to be too 'sentimental'. Few authors, if any, of that time were willing to recognize the serious moral issues which the extirpation of the Great Auk highlighted. Among these, indirectly at least, were the brutalizing conditions of life that existed for many ordinary men and women at the time that the Great Auk's tragedy was being played out, for we should not be unduly surprised to find a degree of correlation between man's — or a man's — behaviour towards other creatures in his environment, and the treatment men mete out to one another.

The widespread misconception about where the principal breeding grounds of the bird had been (which had arisen because all the earlier writers on the Great Auk were Europeans),[1] was inadvertently perpetuated by the compilers of *The Handbook of British Birds* (1938–41) in which reference to the much greater numbers of the species in its New World domicile was omitted. This distorted view was to receive a measure of correction when the widely-read James Fisher came to write the first volumes of his paperback series on *Bird Recognition* after the Second World War. However, although Fisher summarized a mass of information on the Great Auk derived from the earlier writers by stating that the final extinction of the bird came 'after three hundred years of slaughter by seamen for food', he would seem to have failed to recognize the financial motives of the fishermen of south west Iceland and north east Newfoundland in their pursuit of the Great Auk to the point of annihilation.

Until well into the twentieth century, the smaller island communities of the North Atlantic used to kill thousands upon thousands of sea birds annually – and, like the colonists on Newfoundland, did so not only for food. However, this did not have a catastrophic impact on the overall breeding numbers of these birds. This is in contrast to those fishing communities in the New World which ensured that, to use Edward Howe Forbush's phrase,[2] the 'uncounted hosts' of

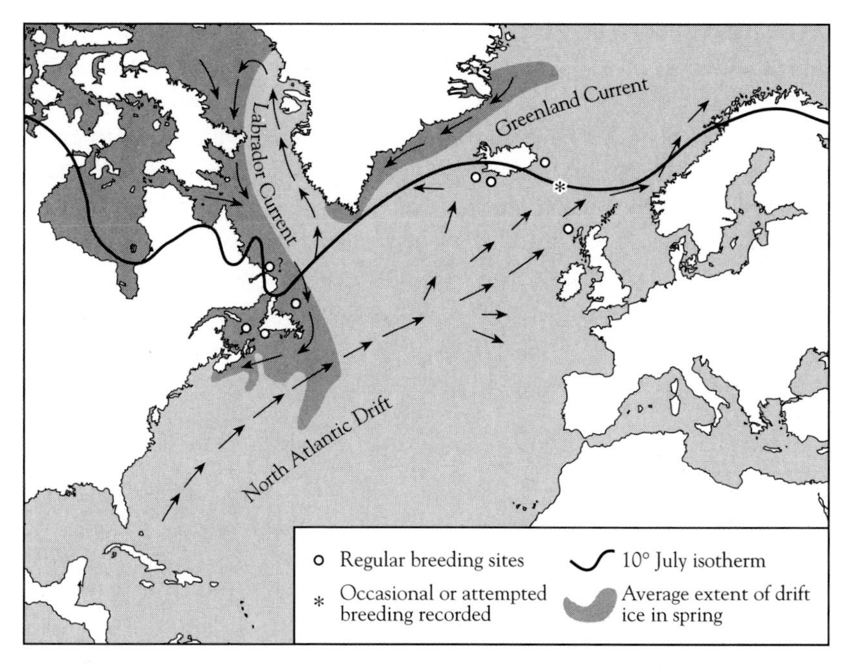

Fig. 2 Map showing the Great Auk's breeding sites in relation to (a) the 10°C July isotherm which marks the ecological boundary between the Arctic and subarctic (Köppen, 1900) and (b) the extent of drift ice during April and May (Nettleship and Evans, 1985). In Europe, breeding commenced in the latter half of May (Martin, 1698; Wolley, 1858); in the New World, at least a month later (Hakluyt, 1600).

Great Auks were reduced to nothingness by men who put their emphasis on short-term gain rather than on the long-term advantage of establishing a perpetually regenerative basis for economic exploitation. To understand how the Great Auk became extinct, we must inquire into the precise circumstances — cultural or economic — which led to the adoption of so obviously unintelligent and counter-productive a policy. Managed differently, the Great Auk would have remained a familiar sight in the busy sea lanes which lie between Newfoundland and the fishing banks.

A full discussion of the destruction of the Great Auk must incorporate a frank account of the uglier aspects of European expansion in earlier centuries — something which the mid-nineteenth-century authors, for whatever reasons, did not emphasize. It remains undeniable that the Great Auk was a victim of the civilization of which we are a part and which too often manifests short-term thinking in its exploitation of natural resources, often in the name of 'progress' or 'development'.

It would be unjust of us to dismiss the authors of the first histories of the Great Auk simply because they were men of their own age, just as we are of ours.

Some of the names which appear in the following pages belong to a small circle of English friends who were among the first to see the need to extend legislative protection to breeding birds and accordingly lobbied Parliament successfully for the passage of the Act for the Preservation of Sea Birds which came on to the statute book in 1869. A number of them in their youth, moreover, had undertaken arduous expeditions in the hope of finding Great Auks at their former breeding stations. No one was to know better than they that extinction was no mere hypothetical possibility. As will become clear, it is no exaggeration to say that the destruction of the Great Auk was the catalyst for the enactment, both by Parliament in the United Kingdom and by the newly established legislatures of its possessions overseas, of some of the first modern laws for the protection of birds.

As British naturalists had been among the first to try to establish the true range and status of the Great Auk, it is appropriate, as it is convenient, to commence this account of the species by finding out just how much British writers knew of it at the beginning of the century during which the last of its kind finally succumbed to the unremitting exploitation which followed the species wherever it came ashore to breed.

Notes

1 Accurate information about the fate of the Great Auk populations in the western Atlantic only first received publication in *The American Naturalist* in 1876.

2 *Birds of Massachusetts and other New England States* (1928).

CHAPTER 1

This rare and noble bird

Until Colonel George Montagu published his *Ornithological Dictionary* in 1802 there existed no conveniently sized volume describing the birdlife of the British Isles. From the publication of this book, which was in marked contrast to the expensive encyclopaedias of previous decades, we can trace the start of an era of popular amateur ornithology which has continued uninterrupted until the present day. Apart from its attractive arrangement for the non-specialist — it was subtitled *An Alphabetical Synopsis of British Birds* — the *Dictionary*'s greatest virtue was that it was based squarely on the most up-to-date information derived from personal observation. Given the limits on travel which then existed, it was not of course possible for any one ornithologist to acquaint himself with more than a fraction of the birds which were to be found further afield than his own county — which in Montagu's case was Devonshire. However, the good conversation and free exchange of ideas among like-minded individuals of that time went far towards making up for any deficiency arising from a lack of first-hand observation.

It may at first seem surprising to a generation familiar with using the term 'Great Auk' for the species which was the Northern Hemisphere's only indigenous flightless bird, but Montagu must be credited with giving wide currency to a name which had been coined only a generation earlier. The eighteenth-century naturalist Thomas Pennant (1726–98) first employed the term in his *British Zoology*, published in 1768. It is arguable, however, that the true inventor of the name should be regarded as the French ornithologist Mathurin Jacques Brisson who, in his six-volume *Ornithologia* published in 1760, designated as *Alca major* what was commonly known as 'Le Grand Pingouin'. 'Great Auk' was a translation (albeit a loose one) of Brisson's new appellation by Pennant.

In Montagu's *Dictionary*, after a conventional description of the plumage based on close examination of one of the few museum specimens then in existence, we may read the following succinct account of the habits of this species:

> The smallness of the wings renders them useless for flight, the longest quill feather not exceeding four inches in length. These, however, are admirably adapted to its mode of life and are of peculiar use in diving under water, where they act as fins; by which means it pursues its prey with astonishing velocity.

Although the sources of this information are not given we may reasonably assume that it was derived from one of the few naturalists, such as Sir Joseph Banks (1743–1820), to have visited the foggy shores of Newfoundland a generation previously when the species was still quite common. Regarding the Great Auk's status as a British bird, Montagu's information was founded on the little that had been published (see Chapter 7) in times past: 'The bird is only found in the most northerly parts of the kingdom; it is said to breed in the isle of St Kilda.'

If a degree of uncertainty regarding the exact range of the species still existed, the reason is not difficult to see. Hitherto, the Great Auk had been known by a bewildering variety of names, some of them in Latin for the purpose of classification whilst others were derived from the vernacular of unlettered seafaring men. Geirfugl,[1] Garfugel, and Gearbhul[2] were used by the descendants of the Norsemen, and Aponar (with variations in spelling), Binocle, Moyack, the Penguin, and, perhaps, the Wobble, by voyagers to the cod-rich seas off North America. As will become clear in due course (see Chapter 6), this multiplicity of names, be they learned or not, was to lead to considerable confusion on the part of those who sought to give a full account of the species. This was further compounded by the Great Auk's superficial resemblance not only to the Great Northern Diver (the Common Loon of North America) but also to the quite unrelated penguins which we associate today with the Antarctic pack ice.

In fact, the Great Auk was the *original* penguin and its name was simply transferred by English and Spanish mariners to the *Spheniscus* penguins of the South Seas. French seamen however distinguished the southerly birds by the name of 'manchots' (on account of their long, sleeve-like wings) and hence in French 'pingouin' is reserved exclusively for two North Atlantic species — the Great Auk and the Razorbill. But things might have turned out otherwise. After visiting the Island of Ascension in the south Atlantic in October 1555, the navigator André Thevet published the following account of a species of penguin, perhaps the Gentoo Penguin, in *Les singularitez de la France antarctique* (1558), given here in the English translation which appeared a decade later:

> Furthermore in this island there is a certain kind of great birds that I have heard called Aponars; they have little wings, and therefore they cannot fly. They are great and high like hernshaws [i.e. herons], the belly white and the back black as coal, the bill like to a cormorant; when they are killed they cry like hogs.[3]

> I have thought good to speak of this bird among others, for that there are found a great number of them in an Island lying toward the Cape of Good Speed [i.e.

Cape Bonavista] on the coast or borders of New-found-land, the which was named the Isle of Aponards.

Great uncertainty about both the affinities and the range of the Great Auk was to persist until well into the nineteenth century,[4] endowing the species with an aura of mystery. The pulses of naturalists and sportsmen in Montagu's day were bound to quicken whenever the bird appeared, at irregular intervals, in north western Europe. The desire to obtain specimens increased just as the number of contemporary reports that the species was becoming extremely scarce (for reasons unknown at the time), also increased. The zeal of collectors ensured that the flightless Auk had almost no chance of surviving once its numbers were reduced to the merest handful by parties of 'mercenary and cruel' men (of whom more anon).

It is not least among the ironies shrouding the history of the Great Auk that those scientifically minded gentlemen who furnished the demand for 'skins', which ensured that a brisk trade continued until the ultimate extinction of the species, are among those who must be counted the bird's greatest admirers. The

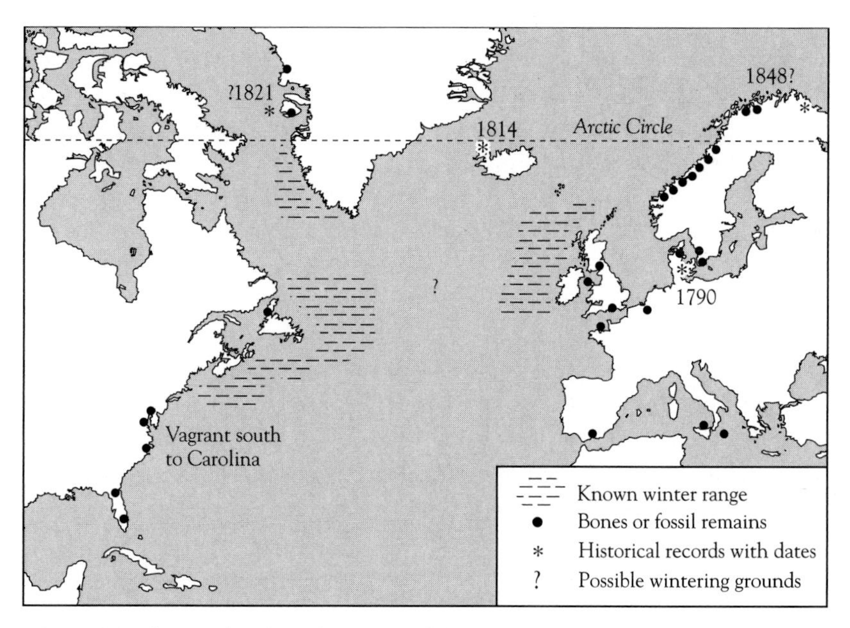

Fig. 1 Sites where archaeological remains of the Great Auk have been found outside the known range of the species in historical times (after Bourne, 1995; Fuller, 1999). Other remains have been found in Nova Scotia, southern Greenland, Iceland, Ireland, and the Outer Hebrides. In general, remains from Western Europe are much older (*c.*5–2500 millennia) than those from Greenland and North America (*c.*0.5–2 millennia). How far the change in distribution of the Great Auk is the result of climatic change, an increase in predators, or of human exploitation remains unclear.

Great Auk is far from being the only species to which collectors have delivered the final blow but there are very few creatures which have been extirpated solely as a result of collecting — an activity which is usually 'the final straw' for one already threatened. In the case of the Great Auk, moreover, the imperfect knowledge of its range meant that when, in the 1840s, Etatsraad (Councillor) D.F. Eschricht acquired two specimens (the last such to be fully authenticated) for the Zoological Museum attached to Copenhagen University, no one had any idea of just how tenuous the bird's hold on existence had become. Almost a decade later Professor William Macgillivray of Marischal College and University, Aberdeen, in volume 5 of *A History of British Birds* (1852), could write:

> This very remarkable bird is an inhabitant of the Arctic seas and in its habits resembles the Razorbill and Guillemots; but its history has not been satisfactorily traced and of its distribution we know only that it extends from the extreme north to the Orkney islands and St Kilda...It appears to be gradually diminishing in numbers, and is generally considered a very scarce bird. It is certainly so as British, for not more than ten are alluded to as having occurred in our seas.

The specimen which was used by many late nineteenth-century illustrators was acquired by Dr William Leach in 1819 on behalf of the British Museum at an auction in Piccadilly of the contents of the Loudon Museum which was the private collection of an energetic naturalist, William Bullock. In his *Companion* to the Loudon Museum, published a few years earlier, Bullock gave a brief account of how the bird came into his hands:

The Great Auk or northern penguin (*Alca impennis*)

> Of this rare and noble bird we have no account of any having been killed on the shores of Britain, except this specimen, for upwards of an hundred years; it was taken at Papa Westray, in Orkney, to the rocks of which it had resorted for several years, in the summer of 1813, and was finely preserved and sent to me by Miss Traill of that island.

A fuller account, contemporary with the above, appeared in the *Scots Magazine* in March 1814 written by Mr (later Dr) Patrick Neill, a knowledgeable naturalist who had visited the Orkneys some years earlier. If accurate in all essentials this is the last record of attempted breeding by this species in the Scottish Isles:

> In the summer of 1812 Mr Bullock, of the Loudon Museum, being in Orkney, was informed at Papa Westray that the King and Queen of the Hawks [*sic*] had of late years frequented the bays of the island in the summer season. He soon after learnt that the female had been killed with a stone while sitting on her egg, and that the male was still in the neighbouring bay. He had the satisfaction of getting sight of him and chased him for several hours unsuccessfully with a six-oared boat. He

> dived most dexterously and made great progress under water, so as effectually to elude his pursuers. These facts are stated by Mr Montague [*sic*] in the appendix to the *Ornithological Dictionary* lately published. We have now to add that last summer the King of the Hawks again returned to his former haunts; but he was solitary having been unsuccessful in procuring another mate. The zeal of the islanders being roused, he was at length killed, and through the attention of the family at Papa Westray, the specimen was transmitted to Mr Bullock. This is the only British specimen known to exist.

The 'King' was shot by a local man, William Foulis, as he rowed southwards along the coast of Papa Westray on his way back from fishing, managing to surprise the bird by keeping close in under the cliffs. The bird was struck as it leapt down into the water on sighting Foulis' boat coming round the headland of the small cove where it had its favourite resting place on a ledge in front of a small cavern on the Auk Craig. We can be confident that this occurred in 1813, probably in the month of May.

Predictably, perhaps, recollections dating from the middle of the nineteenth century of the exact circumstances surrounding this pair of Great Auks on Orkney reveal one or two discrepancies, particularly with regard to the year in which the male bird was shot and whether or not it outlived the female. Fortunately it can be shown that Neill's account is correct for there is no reference to the Great Auk in Bullock's *Companion* for 1813 as there would have been

Fig. 2 East side of Papa Westray. Engraving from *Vertebrate Fauna of the Orkney Islands* (1891) by J.A. Harvie-Brown and T.E. Buckley.

if, as alleged later, the 'King' was killed in 1812. Likewise there is no reference to this event in Montagu's *Appendix,* also published in 1813. Later, in 1858, a nephew of the Miss Traill who had sent William Bullock his specimen wrote that he understood the male bird to have been acquired in 1812 adding, in words which circumstantially confirm Patrick Neill's account of a last, failed, breeding record, 'Some boys or lads afterwards killed the other in the Craig with stones; but it was not got at the time, but I believe drove on shore some time afterwards, but so much decomposed that it was quite useless.' This seems to be nothing more than a simple inversion of the relevant dates.

Although these two birds from the Orkneys are among the last to be recorded in British waters, reports of the occurrence of the species continued almost until it became extinct, a fact which suggests that the bird would have been of more frequent occurrence offshore when it was still plentiful. It is certain, too, that neither Bullock nor Macgillivray was aware of all the records in the public domain,[5] let alone those which went unpublished. In *The Natural History and Antiquities of Northumberland* (1769), privately printed by John Wallis, can be found the following rather quaint entry which remained forgotten for many years: 'The Penguin, a curious and uncommon bird, was taken alive a few years ago in the island of Farn, and presented to the late John William Bacon, Esq. of Etherston with whom it grew so tame and familiar that it would follow him with its body erect to be fed.'

Two years after the auction of the contents of William Bullock's museum, the Revd Dr John Fleming, Minister of Flisk in Fife and author of *The Philosophy of Zoology,* was one of the very few front-rank naturalists to have a live Great Auk pass through his hands. Had Fleming known that he was in fact to be, or even likely to be, the last British scientist to do so, he would surely have left us a fuller account than he did. In August 1821 he accompanied the commissioners of the Northern Lighthouse Board on their annual Hebridean tour of inspection, concluding his cruise at the Mull of Kintyre on the 26th of the month. About a week earlier they had left Scalpay (or 'Glass') near Harris, and in his account of the voyage, published in the *Edinburgh Philosophical Journal* in 1824, he wrote as follows:

> When on the eve of our departure from this island, we got on board a live example of the Great Auk (*Alca impennis*) which Mr Maclellan, the tacksman [i.e. tenant] of Glass had captured some time before off St Kilda. It was emaciated, and had the appearance of being sickly, but, in the course of a few days, it became sprightly, having been plentifully supplied with fresh fish, and permitted occasionally to sport in the water with a cord fastened to one of its legs to prevent escape. Even in this state of restraint it performed the motions of diving and swimming under water with a rapidity that set all pursuit from a boat at defiance. A few white feathers were at this time making their appearance on the sides of its neck and throat, which increased considerably during the following week and left no room to doubt that, like its congeners, the blackness of the throat feathers in

> summer is exchanged for white during the winter season. I may add that the black colour of the throat of the Razor-bill (*Alca torda*) was at this time undergoing a similar change.

Fleming published the measurements,[6] together with a few more details of this bird, in *A History of British Animals* (1828): 'Legs black. Irides chestnut; margin of the eyelid black. Inside of the mouth orange. Head, back, and neck black, the latter with a brownish tinge.' These few details are highly significant — firstly, for confirming the iris colour of the species and, secondly, for including the only reference based on personal observation to the colour of the gape. This serves to underline how little was known of the Great Auk before it was lost for ever. Fleming added one extra detail of its behaviour which is consistent with other accounts of the bird in captivity: 'When fed, in confinement, it holds up its head, expressing its anxiety [i.e. eagerness] by shaking its head and neck and uttering a gurgling noise.'

This last piece of information is one of only two or three descriptions to come down to us of the kind of noise that the Great Auk made. As to the circumstances in which Fleming's bird was first captured, it was a good many years before details came to light. In June 1880, Robert Gray, who had published *Birds of the West of Scotland* in 1871, received the following information from a Mr R. Scot Skirving:

> I think you will be interested in knowing that when on St Kilda on the 14th of this month, I found there was a man still living there who assisted in the capture of Fleming's Great Auk in 1821–22.[7]
>
> Having shown a drawing of the Auk to the collected natives to see if they had any knowledge of it, they said they knew it used to be there long ago, but they had never seen it. Subsequently they told me the man was still there who 'caught the last Great Auk'. I had him immediately brought to me. His name is Donald McQueen, Sen[ior]. He is a very little man, and is also so much bent, that he does not stand much higher than the Great Auk did. He said he was 73 years of age but to all appearances he is considerably more. Donald disclaimed having been (as his nephew reported) the person who actually caught the Auk. He informed me that he was one of four persons in a boat on the east side of the island when they discovered the bird sitting on a low ledge of the cliff. Two of their number (then young men) were landed, one on either side of the bird, and at the same distance from it. These two cautiously approached it whilst he and another boy rowed the boat straight towards the Auk which ultimately leapt down towards the sea, when one of the youths, having got directly under it, caught it in his arms. The old man with much animation went through the pantomime of grasping a supposed bird in his arms and holding it tightly to his breast.
>
> A partial error of the old St Kildian [*sic*] served to identify the Auk with Fleming's. He said the people who got it 'tied a string to its leg and killed it.' When I told him

> they did not kill it, he said he might have forgotten what he heard about it after it was taken away.

The bird was in fact intended for a museum in Edinburgh, but managed to escape from the lighthouse-keeper at Pladda as Dr Fleming's sea trip drew to its close. It must have been only a matter of time before the bird, with the long cord still attached to one leg, became ensnared and drowned.[8]

The accounts of both Bullock's and Fleming's Great Auks, the best known examples from British waters in the early nineteenth century, reveal that, after suffering for so long from a shortage of information about the species, naturalists were collating and publishing what accurate data they could. Such endeavour was not confined to the British side of the English Channel and continental ornithologists may be regarded as, in certain respects at least, better informed as to the status of what was, by then, no more than a relict species.

Notes

1 The second *g* in the Icelandic name Geirfugl (meaning 'spear-bird', after the shape of the bill) is so soft as to be almost unpronounced.

2 The *bh* is pronounced as *v*.

3 It is not certain whether this description is based on the Great Auk as described to Thevet some years earlier or on his own observations of whatever species of *Spheniscus* penguin bred on Ascension Island in the sixteenth century. Certainly the cormorant-like beak suggests the former, but the noise he describes them as making when being clubbed is remarkably similar to the description given in 1823 of a penguin 'hunt' in the Falkland Islands by René Lesson. South Sea penguins were, for a time, over-exploited to furnish bait for the crayfish industry: see Cherry Kearton's *Island of Penguins* (1930). There is no evidence indicating whether visiting sailors or some introduced creature, such as the rat, is responsible for having extirpated the breeding population of penguins on Ascension Island. W.R.P. Bourne has expressed, *in litt.* to the present writer, his doubts regarding the suitability of Ascension Island as a breeding site for any of the Southern Hemisphere penguins. It remains certain, however, that Great Auks and *Spheniscus* penguins were initially regarded as conspecific.

4 As late as 1865, Professor R. Owen in his 'Description of the Skeleton of the Great Auk' published in the *Transactions of the Zoological Society* was at pains to demonstrate that the Great Auk was unrelated to the *Spheniscus* penguins. This was probably unnecessary for his immediate readers but the mere fact that he felt obliged to do so illustrates how persistent popular misconceptions can be.

5 There is, for instance, a well-executed if somewhat childish drawing of the Great Auk dating to *c.*1652 among the Additional MSS in the British Library. The bird is depicted in an unnatural posture suggesting that it was based on an early mounted specimen but beneath it in a carefully written hand are the words: 'These kind of birds are about the Isle of Man.' The drawing is attributed to one Daniel King who illustrated *A Short Treatise of the Isle of Man* for James Chaloner, an ardent Cromwellian, who was appointed as a commissioner of the island by Thomas, Lord Fairfax, in December 1651, after the fall of the island to Parliamentary forces in the Civil War.

6 These are not significantly different from those given by Audubon and Macgillivray (see Appendix 1) except for the length. Fleming gave this as fully 36 inches (presumably taken when the bird was at its most attenuated) in contrast to Macgillivray's measurement of 25 inches for a mounted specimen.

7 The true date is 1821 but as Fleming gave the differing dates in his two accounts of the Auk, for many years there was confusion over this point. The matter was not resolved until the antiquarian J.A. Smith, MD, wrote to the Northern Lighthouse Board to enquire in 1880.

8 In his *Birds of the West of Scotland* (1871) Robert Gray referred to a handwritten list of the birds of Renfrewshire in which the Great Auk was mentioned, further enquiry by him revealing that one was said to have been washed ashore at Gourock some fifty years previously. As the Great Northern Diver — 'a species which is sometimes confused with it in districts where Gaelic names only are in use' — was already on the list, Gray suggested that the Gourock record may 'account for the fate of Dr Fleming's specimen, Gourock being situated at a part of the Firth of Clyde likely to be visited by a crippled bird from the coast of Arran, where this half emaciated Garefowl regained its freedom'.

CHAPTER 2

The Icelandic bird skerries

On a windless day in the late[1] summer of 1813 an armed schooner, the *Faeroe*, lay becalmed off the chain of volcanic islets known as the 'Fuglasker', or 'bird skerries', far out to sea off south west Iceland's Reykjanes promontory. The ship had been sent as a matter of urgency to obtain supplies from Reykjavik for the half-famished population of the Faeroe Islands. Denmark, which ruled both the Faeroes and Iceland at this time, was a satellite of Napoleonic France and in consequence was in a state of war with Great Britain. Having no particular quarrel with the peaceable inhabitants of Denmark's dependent territories, the Government in London, at the instigation of the President of the Royal Society, Sir Joseph Banks, had recently decreed that they should no longer be subjected to molestation and had permitted the resumption of trade. In Copenhagen, however, all commerce with a hostile state was regarded as a form of collaboration which merited the customary penalty, with the predictable result that the islanders, obedient, either through fear or inclination, to their Government, lived in greatly impoverished circumstances.

Lying off the bird skerries, the *Faeroe*'s commander, Peter Hansen, would have recalled his previous visit five years earlier. Then he had been pressed into service by a party of free-booting British privateers who, having sacked and looted the Faeroese capital, Thorshaven, were in need of a local man to guide them to Iceland for a similar purpose. Some days later, after letting the people of Reykjavik taste the horrors of war, the crew had spent a whole day on the skerries killing many birds and, we are told, destroying their half-fledged broods. This misdeed was carried out with the express intention of denying to the villages of the Reykjanes promontory any chance of supplementing their diet from this source. Sea bird colonies were to take so much of the brunt of the economic distress occasioned by war that it is fair to say that the age-old rivalry between England and France, which reached its greatest intensity in the eighteenth cen-

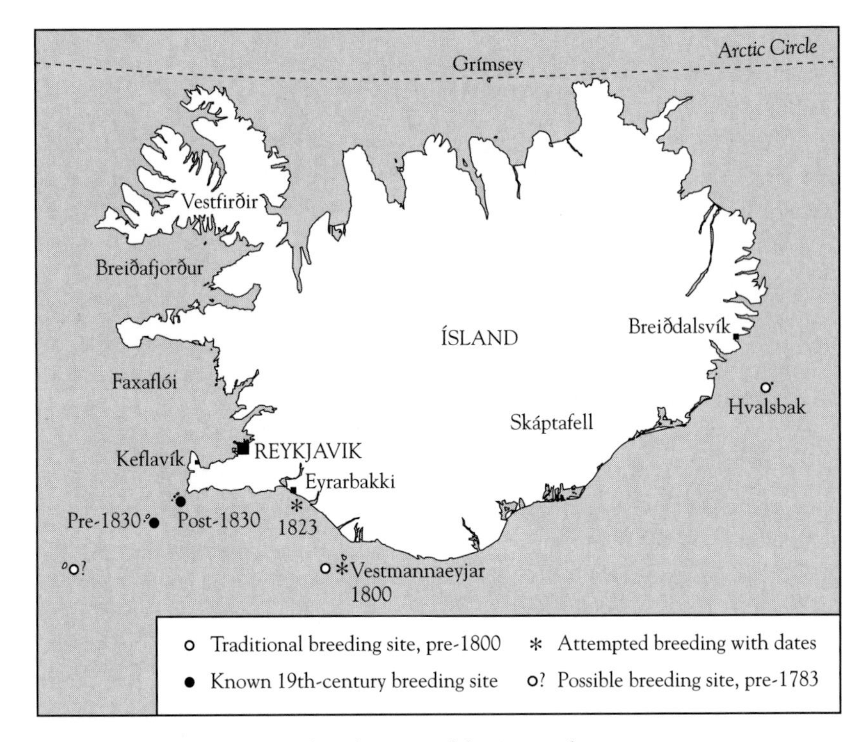

Fig. 1 Map of Iceland showing breeding sites of the Great Auk.

tury, was a significant contributory factor to the depletion in numbers of all manner of sea fowl, especially members of the Auk family.

It is doubtful that Peter Hansen saw any Great Auks — or 'Geirfuglar' as they are known to the people of Iceland — on his visit to the skerries in 1808. By late July the parent birds would, weeks before, have led away their young to the safety of the deep; however if any, young or old, had in the meantime grown accustomed to rest up for part of the day out of the water, then conceivably they too might have been engulfed in the local catastrophe which overtook the Geirfuglasker (Gairfowlskerry). This was the only island in the chain on which the Great Auk could conveniently come ashore. Only a generation before there had been another island, the Blindefuglasker (the 'Sunken bird skerry') which, like the Geirfuglasker itself, was a relic of the rim of the crater of an ancient volcano of enormous size. This site, too, might have been suitable for the Great Auk to breed on. It had, however, disappeared in 1783 in a complex series of eruptions centred on the Skápta-jokull in the south of the mainland. The consequence had been a devastating environmental catastrophe in which many lives were lost — the Icelandic economy, then as now, being largely agrarian. Noxious gases had killed livestock by the thousand and, effectively turning the rain into sulphuric acid, had destroyed both grazing and hay crops. Even the fishing had been affected, for the force of the eruptions had been so severe as to leave the seas off the south and west coasts of the country carpeted with a thick layer of pumice.

In 1813, as Peter Hansen's crew, weary of a second manmade disaster to afflict the Faeroese within the space of five years, prepared to visit the Geirfuglasker on 24 August (or, as seems more likely, July) 1813, they must have considered the flat, calm water a heaven-sent opportunity to obtain provisions, which could be exchanged for essential commodities once they arrived in Reykjavik. Forty-five years later one of them, Daniel Joensen, recalled the events which had occurred that day:

> [F]ive of the men got into the boat, Skipper Hansen and C. Hansen remaining in charge of the schooner, and pulled to some *drangs* [i.e. islets or stacks], and landed on the largest they saw...they could not climb up to the top of this *drang*, but they succeeded in procuring a few Gannets but no Guillemots. Just as they

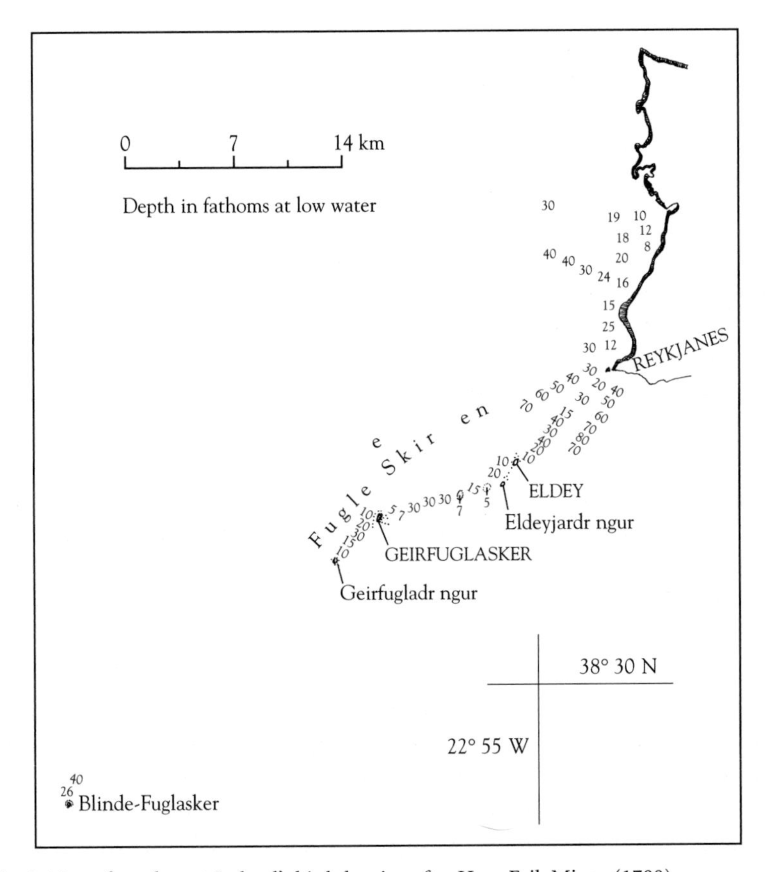

Fig. 2 Map of south west Iceland's bird skerries, after Hans Erik Minor (1788).

> were returning on board, the tide drifted the schooner from them, and then they noticed a lot of birds on a skerry with a flat top: this skerry was further from the shore than the *drang* they had already visited. They pulled their boat to this skerry; at the east or north east end was a ledge, and as they approached it they saw that it was covered with Gare-fowls. The weather was fine and clear at the time. When they got close to the ledge many of the Gare-fowls shuffled into the water and swam away: the boat was kept with its stern to the ledge and backed in on the swell of a wave; Daniel Joensen, Paul Medjord and H. Joensen sprang on to the ledge and seized the remaining gare-fowls wth their hands, wringing their necks. Daniel Joensen does not remember whether they took fourteen Gare-fowls or eleven, but either one or the other number; they saw neither young nor eggs on this ledge, and as the wash of the sea broke over it, it would have been impossible for eggs or young to have remained there. Many more Gare-fowls went to sea than what they killed. The Gare-fowls were only on the ledge, and no other birds were seated alongside of them.

The three men proceeded, with a little difficulty, to scale the skerry, finding Gannets and Guillemots on the upper reaches. They concentrated their energies on the latter, killing about two hundred before a fog came down, hiding the schooner, just as the tide began to turn. They decided to beat a hasty retreat and then found that 'the boat was so full of birds they had to leave some of the dead Guillemots on the ledge, but they did not leave a single dead Gare-fowl behind them'. Daniel Joensen, his namesake H. Joensen, and Paul Medjord had been able to kill such a large number of birds with their bare hands chiefly because, usually unmolested, the 'Geirfuglarnir' showed no fear of strangers.

Until the destructive raid of 1808, the bird skerries had enjoyed many breeding seasons free from serious disturbance, the Icelanders of Cape Reykjanes finding the benefits of a trip to the islands not worth the risk to life and limb. Moreover, securing the birds themselves does not seem to have been the objective of such expeditions even when they had been a regular feature of the summer season, but, rather, the Auks' eggs. Niels Horrebov (or Horrebow), author of *The Natural History of Iceland* (1752, English edition 1758), wrote concerning the 'Geyervogel' as follows: 'The inhabitants at a certain season go to these islands, though the expedition is very dangerous, to seek after the eggs of this bird of which they bring home a cargo big enough for eight men to row.'

There is a degree of exaggeration in this for we should not expect those who made the journey to restrict themselves exclusively to the eggs of just one species, but to take any that came within reach. Additionally, the fact that eight-oared boats were used is less a reflection of the quantity of eggs that could be gathered and more an indication of the dangers of the roost, as the meeting of contrary currents off Cape Reykjanes can prove, especially in conjunction with a running tide, exceptionally difficult to negotiate for even the sturdiest of boats. Horrebov continued: 'The danger and difficulty consists of getting ashore near these cliffs which lie six or eight leagues out at sea where the water generally runs so high that, if the boat be not very carefully managed, it runs a risk of being dashed to pieces by the waves.'

In 1639, two boats out of four which made the expedition were lost, the crews of the remaining two having to make space for their companions and struggle back to Cape Reykjanes as best they could; in 1628, disaster had struck when twelve villagers lost their lives on such a trip. Not surprisingly subsequent generations were filled with trepidation at the prospect of landing on the skerries. Later in the same century a local Lutheran pastor coined a grim little couplet[2] which translates:

> I have never trusted myself to go to Geirfuglasker
> As, on account of the surf, boats were broken by the waves there.

Nevertheless, as we can infer from Horrebov's account, a few intrepid individuals made the trip from time to time, although we may be justified in asking whether this was done quite as regularly as this writer suggests. An early eighteenth-century account of the Geirfuglasker, attributed to Guðni Sigurðsson (a copy of which is retained by the National Library in Reykjavik), describes the Garefowl as being there 'not nearly so much as people suppose' and, in an attached sketch, sixty-one birds are shown at one of two suitable breeding areas on the island. A footnote, probably written by the transcriber and so dating to *c.* 1770, gives a conventional description of the bird adding that it yielded '1/2 pd.' (250g) of fat and that its meat tasted quite good. The same writer gave his opinion that, from a human point of view, all the feathers apart from those on the neck were quite useless because their shafts were so stiff that they pierced 'every coverlet and even Dutch sail-cloth.' The Great Auk's egg is accurately described as being as big as a swan's except that, like those of other sea birds, it tapered at one end. Regarding their colour, we read that no two are alike, being 'artistically patterned with different colours,' leading the writer to conclude: 'in short: Nature has here produced a masterpiece. I have known Danes give 8 or 10 skildinga (shillings) for a blown egg. *Rara avis in terris.* [It is a rare bird on land.]'

Horrebov himself went some way, in a footnote, towards confirming the general impression, conveyed by these details, that the Great Auk on its Icelandic breeding grounds was known to comparatively few people. Citing Johan Andersson, the author of *News from Iceland, Greenland and the Davis Strait* (1747), he stated that the Garefowl:

> is not often seen here except on a few cliffs to the west and that the Icelanders, naturally superstitious, have a notion that when this bird appears it portends some extraordinary event. Of this he assures us of his being told that [in 1729] the year before the late King Frederick IV died, there appeared several and that none had been seen before for many years.

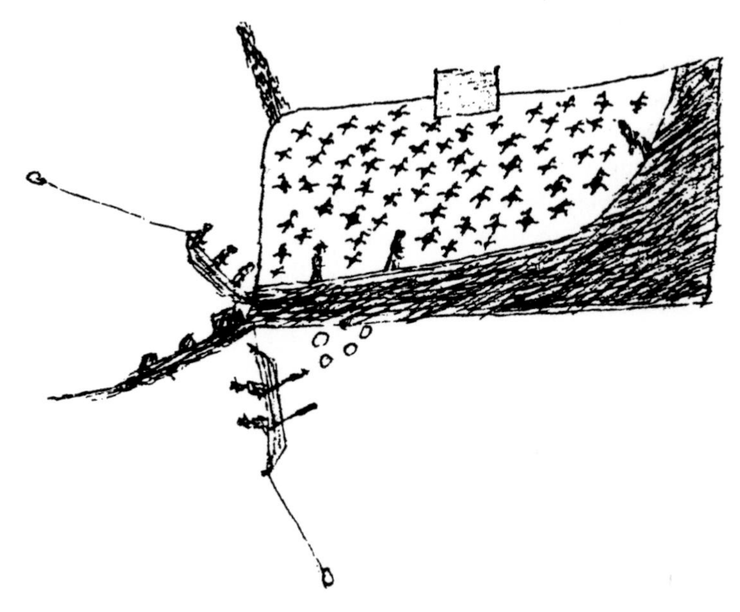

Fig. 3 An eighteenth-century sketch, preserved in the National Library of Iceland, showing the view southwards along the east coast of the Geirfuglasker, with boats — anchored with rocks — waiting at the two landing places.

Horrebov's own conclusion is somewhat less dramatic: 'Though there are not so many of these birds as of other seabirds, yet they are not scarce. They are frequently seen and those that go to take their eggs from them see enough of them. The eggs are very large and almost as big as ostriches' eggs.'

We can assume from the last remark that the Great Auk was known to the majority of Icelanders more by reputation than by actual firsthand experience and so infer that even on these volcanic skerries, which were soon to be established in the minds of European ornithologists as the *locus classicus* of the species, the occurrence of the bird was unpredictable, its numbers liable to fluctuation, and its appearances and disappearances sufficiently mysterious as to render it a bird of ill omen.

The raid on the skerries carried out in August 1813 inflicted a severe blow on the modest numbers of Great Auks which frequented these remote islets. Conceivably, despite the apparent clarity of Daniel Joensen's recollections, the toll was even higher: his crew-mate Paul Medjord, who likewise gave an oral deposition in 1858, insisted that the number of Garefowl conveyed to Reykjavik had been no fewer than twenty-four. There is no way of being sure whose recollection was the more accurate, but we may wonder whether the birds would have stayed on the ledge long enough for three men to dispatch an average of eight apiece. Whichever total is correct it is certain that the figure of 'fifty or sixty' given to the British ornithologist John Wolley when he visited the Faeroes in 1849 was, as Wolley himself later realized, based on a misapprehension.

Given the extremities to which warfare had driven them, and their lack of knowledge of the status of the Great Auk at that time, the poor Faeroese should not be blamed unduly. As this was probably the only significant blow inflicted on the Great Auk population in Iceland between 1800 and the submergence of the Geirfuglasker in 1830, after which the species became more vulnerable to collectors, it was one which the small local population could probably have sustained.

From about Horrebov's time, the Danish fishermen who visited Icelandic waters came to know the waters around the Fuglasker better — and saw the Great Auk more often — than the inhabitants of Cape Reykjanes, and in consequence the islands acquired Danish names in addition to their Icelandic ones. These new names were no more than aids to navigation, the different islets being described according to their shapes: one, a small, flat-topped 'drang' (perhaps the one visited by Daniel Joensen and Paul Medjord in 1813) reminded the newcomers of a soldier's cap; while the Geirfuglasker itself suggested a barn, giving rise to the name 'Ladegaarden'. By contrast, the Icelandic names, such as Eldey (meaning 'Fire Island'), often reflected the volcanic activity for which the region is notorious. Eldey, the most precipitous of the skerries and, as such, not the Great Auk's first choice as a breeding site, lies about

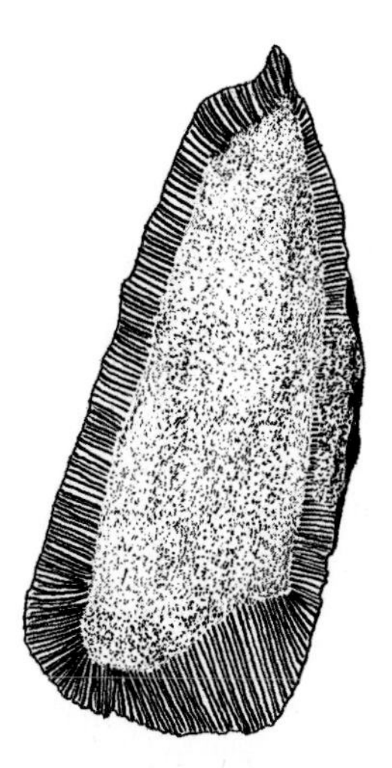

Fig. 4 The Geirfuglasker, 'a vegetated rocky fastness...in size comparable to a typical cow pasture'; the cliffs 'sheer all round except at the highest point of the eastern most side where both of the access points ('uppgöngurnar') are'. From H.E. Minor's map of the west coast of Iceland (1788) and an eighteenth-century description attributable to Guðni Sigurðsson.

thirteen miles (20.9 km) from Cape Reykjanes — in other words, about halfway out to the site of Geirfuglasker. On account of its rotund shape and whitish appearance (a consequence of the deposits of sea birds being washed down the cliff faces) it was referred to by the Danish fishermen as the 'Meel-sækken' (in English, the 'Mealsack').

The first European ornithologist to undertake a visit to the breeding grounds of the Great Auk was a Danish regimental quartermaster, Friedrich Faber, who was based at Eyrarbakki on the south coast of Iceland only a short distance to the east of the Reykjanes promontory. His account of the trip appeared in the German ornithological journal *Isis* in 1827. The difficulties and disappointments which would attend an expedition to those surf-bound, barren islands can be fully appreciated in his narrative:

Voyage to the Bird-skerries

Late in 1820 I travelled to Reykjanes in order to hire a boat for the Bird-skerries the following summer; I myself was not able to get hold of one because the Icelanders are reluctant to risk going out there in boats, so I commissioned a skilful man, a certain farmer Viljamsön; this did not succeed either, so I left my winter quarters at Örebacke on the 24th of May 1821 and travelled again to Reykjanes to visit the appropriate organizations and awaited a convenient opportunity, because it is only when the north wind renders the waves round the islands of no great consequence that they can be successfully negotiated. Often it is as much as a whole month before wind and weather become favourable. In selecting the month of June for this boat trip I had good reason because then, when the weather is calmest, is the best time for all the members of the Auk family to breed and they are certain to be met with at their nests. Meanwhile two Danish travellers came to Iceland on the 2nd of June, the Count von Raben and the Chancellery official Morck, whom I met in Reykjavik: we made an agreement to sail to the Bird-skerries together. On the 13th of June I proceeded to the commercial district at Keblavik, where the other travellers arrived on the 26th of the same month. It was advisable to hire a small ketch for the cruise, one which could undertake a journey lasting several days, for it was by no means certain that one could simply head out to the islands and hence it was necessary to take bedding with us: anyway a boat was not available. We hired a ketch there for 20 rix-dollars, whereupon a stiff southerly wind blew up creating some powerful breakers. Thus the hunt was called off once again and the whole undertaking went back to being threatened; frankly neither was a good omen at this time and at first I had no confidence of a satisfactory outcome. Then the wind went round to the north again and we immediately sought to take possession of the ketch; but the owner[3] now demanded 32 rix-dollars, or 40 Thalers — a highly significant price for a sea-trip of not many days and of no great distance; nevertheless we paid it, and hoped by means of an outright defeat to take the harmless *Alca impennis*; at the same time the boatman stipulated that if we obtained as many Great Auks as we wanted, then he should be allowed to fill his boat with whatever remained, something which we, of course, freely gave him permission to do.

At midday on the 29th we all embarked and so met up on the biggest ketch. Two [Danish] miles south west of Iceland's Reykjanes promontory lies the first of three bird-skerries, Eldey or *Meelsäkken*, i.e. the Mealsack as it is called, a high, broad, and inaccessible rock, only frequented by *Sula alba*; two [Danish] miles from this lies the supposed residence of the Great Auks, the Geirfuglasker, a low-lying, broad island, accessible on the west side, and another half a [Danish] mile further out to sea, the third and last bird-skerry — *Grenadeerhuen*, Grenadier's Cap, on account of its narrow, high, steep-sided aspect. We struggled against a head wind and powerful surge; towards evening, for the first time on one of my Icelandic trips, I went down with that vile and revolting thing, sea-sickness; so severe was it that when I was meant to climb from the small yacht into a bigger one that came alongside us, I had to summon all my strength. The blame for these misfortunes I lay on my daily consumption of milk which formed my staple diet during my two year stay in Iceland and, according to the opinion of several Icelanders, it is indeed supposed to promote sea-sickness; also being in close proximity to the awful stench of fish in the small boat bears at least some of the responsibility; however, after sleeping for a night in the outrigger, I awoke freed from this loathsome thing.

The next day, the 30th of June, about midday, we had reached the furthest island, the Grenadier's Cap, where we, however, did not expect to find our bird for the simple reason that it is insurmountable for a flightless fowl; all that was breeding on top of it were some Gannets and foolish Guillemots, and countless Fulmars flew effortlessly round our ketch. The powerful breakers on this island did not bode well for the main purpose of our expedition. In the evening we sailed for our destination, the Geirfugl-skerry, and crossed over there during a night as bright as day. On the water drifted sleeping white Gannets, which, heads tucked under wings, looked just like shuttlecocks. When we neared the island I searched hither and thither longingly to detect a swimming Great Auk, but without success.

Next morning at 3 a.m., as the sea had become calm, a landing on the skerry was decided upon, and as there was only room in the small skiff for one person, apart from the lad engaged for the rowing, so the Count, an energetic and thrusting man, was the first in accordance with his wishes. Once the boat had set off, the ketch tacked back and forth at a distance of a couple of gunshots from the island since naturally we could not lower the anchor in the open sea. Admittedly the Count encountered some difficulties with the breakers on the island, but these were not so powerful that one could not risk setting a course through; he gave the command to the Icelander who, it has to be said, at first refused, but notwithstanding this, so it was the Count got his opportunity to spring on to a boulder on that shore; meanwhile another wave broke over it and pulled him with it out into the open sea where, if he had not been able to swim, he would inevitably have drowned; he swam back towards the skerry, seized his gun which he had lost at the moment when the surf retreated from the rocks, and climbed the island, but without seeing a single Auk. He regained the boat only with difficulty and over the course of an hour and a half, dripping wet from head to toe, made it back to the ketch where he was glad to get to bed. Then we sailed back to the Grenadier's Cap, scanning the surface of the sea for the Auk, but in vain; it was now so calm that we were able to lie close in under the island.

About noon the same day we described an arc out into the sea and set course once more for the Geirfuglasker; there the sea was calm so the Count and I went in the skiff, while the ketch tacked nearby, and rowed several times round the skerry. We saw every bird on the lower reaches; it had a white and black chequered appearance from the backs of breeding *Sula alba* and *Uria troile* [Gannets and Guillemots] which were sitting in between each other there; but the bird we so anxiously sought was not there. I can unhesitatingly declare, which I would not otherwise be able to assert, that this their best known breeding place in the North, had been totally deserted by them. Perhaps someone else may have better luck; it is possible that they used it for nesting much earlier and that they had already left the breeding site with their young. If the wind had not been so unfavourable we would have been out there two weeks earlier. We shot fourteen specimens of the white Gannet, which flew off overhead, in the space of half an hour; the breeding Guillemots on the skerry did not move from the place on account of the firing.

On the skerry a pair of big, shy seacows — called by Olafsen *Utselur* or *Utskersselur*, by Fabricius *Phoca grypus* and by Neurn *Halichoerus griseus* — had their territory. Disturbed on account of this rare visit, they swam with wild eyes and stiff snouts suspiciously round our boat, that monstrous head always turned towards you. This is Iceland's biggest seacow which always keeps to the most remote islands, shy of the company of its own kind; it is just as reclusive in its dealings with *Phoca vitulina* [the seal] and lives only with its mate. Each pair takes a different district. When we fired at them with lead shot it had no effect.

After a couple of hours we rowed back to the ketch and because we were unable to attain the goal of our cruise, and as continued waiting was of no use, we let the ketch sail back at 8 p.m., and so it was that we passed by that prominent island, the Mealsack. While the ketch was making its crossing, the fishing gear was put out and a significant quantity of cod caught. The wind was fair; the next day, on the morning of the 2nd of July, we were close to the bird-rock at Keblavik and at about 8 a.m. sailed into the harbour of the commercial district after an absence of close on three days. The shore was lined with people who had come out to welcome us back and to see the rare Great Auk; but we could show them nothing but white Gannets. This is my only excursion off the Icelandic coast which did not produce the desired result.

Friedrich Faber had one more chance. According to Erichsen and Schonning's detailed map reproduced in Eggert Olafsen's *Journey through Iceland* (1772) there was another Geirfuglasker lying south west of the Westmann Islands off Iceland's southern coast. Reluctant to go to further expense so late in the season, Faber spent the rest of his leave studying the birdlife of the Westmann Islands and asking the inhabitants what they knew of the Great Auk. One person told him that the bird was seen offshore every spring and added that, as well as being flightless, the bird was blind on account of a large flap of skin over each eye. It must have seemed to Faber that such a misapprehension indicated that the Westmann Islanders had only a very hazy understanding of the species. Eventually, on finding the man who had most to do with fowling operations, Faber learnt that about twenty years previously he had killed an incubating bird discovered low down on the local cliffs, but that this was the only one he had ever seen. From this Faber would have concluded that a future expedition to the Geirfuglasker just out of sight beyond the horizon would have been a pointless exercise. Pessimistic about the status of the Great Auk in Iceland, he returned to his military duties and, when the nights grew long and the days short, commenced his introductory *Prodromus* on Icelandic ornithology. In this little book, published the following year, he reflected on the events of 1813 of which he had heard report when in Reykjavik. 'This, I fear,' he wrote, 'has driven the bird away from the skerry entirely.'

It is difficult not to infer that Faber's judgement was affected by his failure to bring his long-awaited expedition to a successful conclusion. The number of Great Auks taken by Icelanders over the next decade or so — once it was realized there was a lucrative market for them — indicates that the species was far from having been driven away entirely. No one was to know, in Faber's day, that these birds represented virtually all that was left of a once numerous species. It is particularly regrettable that Faber was unable to obtain accurate information about the Westmann Islands' Geirfuglasker, for if a bird had been taken on its egg on the main archipelago as recently as the turn of the century, it is possible that the isolated Geirfuglasker was used by the Great Auk for nesting purposes much later than either Faber or any of the subsequent nineteenth-century commentators suspected.

Once Faber had taken up a new posting in Denmark, he discussed the status of the Great Auk in Iceland, as he perceived it, with other leading Scandinavian ornithologists. Three years before Faber's account of his trip to the bird skerries was published, the ornithologist F. Benicken, well-known in the Royal Danish Museum in Copenhagen, wrote a brief account of the species in the *Isis*. He stated, citing Faber as his authority, that the Great Auk was now 'ausgerotten' (exterminated) on its Icelandic breeding grounds and that the species should be struck off (auszustreichen) the European list. Even if premature, these words should have put ornithologists upon notice that the bird could not tolerate unrelenting interference. In terms which now seem almost prophetic, Benicken wrote that, in view of the slow rate of reproduction of the species, 'through

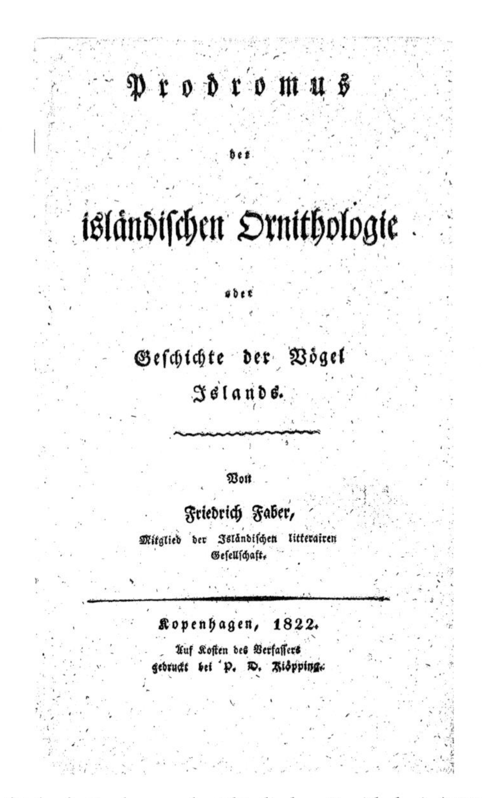

Prodromus

der

isländischen Ornithologie

oder

Geschichte der Vögel
Islands.

Von

Friedrich Faber,
Mitglied der Isländischen litterairen
Gesellschaft.

Kopenhagen, 1822.
Auf Kosten des Verfassers
gedruckt bei P. D. Kiöpping.

Fig. 5 Title page of Faber's *Prodromus der Isländischen Ornithologie* (1822).

killing the old birds at their nests, that it could easily become exterminated is very understandable.' It would not dawn on European ornithologists for another generation that these words, unknown to their author, encapsulated the catastrophe which had overtaken the Great Auk many miles away on the other side of the Atlantic.

Friedrich Faber himself, while appearing to admit that others might have better luck than he did in searching for the Great Auk off Iceland, in fact continued to entertain the gravest doubts as to the species' continued survival. In his most substantial book *Ueber das Leben der Hochnordischen Vögel* ('On the Birdlife of the Far North') of 1826, he could only bring himself to give the unique bird, which he had sought but failed to find, the briefest of entries. His judgement was that its flightlessness was the cause of the downfall ('der Fall') of the entire species.

Notes

1 The two accounts on which the narrative of these events is based differ in a number of points including whether it was the month of July or August that

the expedition was made. In Iceland young Guillemots drop off the nesting ledges into the water in mid-July. Possibly the breeding season may be extended if the first eggs are removed by fowlers, but late August would seem very late indeed for Auks to be frequenting their nest sites.

2 'Ég get ekki gefið mig í Geirfuglasker / eggið brýtur báran pví brimið er.' (Séra Hallkiell Stephansson, *fl.* 1665–97)

3 Jón Danielsson, father of Ketil Jonsson and grandfather of Ketil Ketilsson, both of whom took part in fowling expeditions to Eldey. The last took an active part in the raid in which what are often believed to have been the very last Great Auks of all were captured.

CHAPTER 3

Travels with Audubon in Labrador

For what must have seemed a long but rewarding decade commencing in 1827, the artist and ornithologist John James Audubon worked tirelessly on his illustrations for *The Birds of America*. In this work he combined his own magnificent plates with a lively text which reflected a growing trend towards good field ornithology in the United States. The generosity with which ornithologists of his generation ensured that Audubon had access to specimens of new species as they were discovered is much to be admired. Almost as well known in ornithological circles in Great Britain as he was in his native country, Audubon crossed and recrossed the Atlantic during this decade, visiting friends, borrowing specimens, and persuading those wealthy enough to do so, to support his monumental undertaking by taking out subscriptions. When he finally laid down his brush and pen in June 1838, he had completed 435 colour plates supplemented by over a thousand monochrome engravings and detailed plumage descriptions by his Scottish friend, William Macgillivray.

Despite his prodigious output of work Audubon remained at heart a field ornithologist. In 1833 he had decided to visit Labrador in search of new specimens and to visit *en route* the immense sea bird colonies of the Gulf of St Lawrence. On the 22nd of May that year Audubon wrote to his wife, Lucy, from Eastport, Maine, about the good things he had been hearing from the fishermen of the regions which he would explore for the next three months along with his son, John Woodhouse Audubon and a youthful friend of the family, Thomas Lincoln.

> Had the weather been fair, our little voyage to the Seal Islands might have proved valuable as we hear of many birds breeding there, but as it is it would be folly to attempt navigating through the fog which is so thick that, to use an expression common here, 'one may drive a nail in it and hang his hat upon it'.
>
> The descriptions I have had of the cod-fishing at Labrador border on the marvellous — some owners of vessels who have [been] there several years or, better, say seasons, have told me that in shallow water, say 7 or 8 feet, the cods are

> so aboundant that they appear to touch each other, and that two men have been known to haul up three thousand six hundred in twenty one hours, which is the time they fish per day — these men sleep only three to four hours out of the twenty four — all the time they are on that coast they eat no fish in well-regulated vessels; Beef, Pork, *birds' eggs* and potatoes; with molasses and *water* forms all their food and drink — they are ruddy, strong and as active-looking [a] set of men as can be imagined. Never sick I am told — the climate is so pure, clear and regular and without fogs that they all return home in better condition than that in which they depart from it...We are likely to see between five and six hundred vessels fishing from all nations — One or two British armed vessels are generally there to maintain peace and good faith between the parties engaged — seals are caught in netts by hundreds at a haul etc. etc. — All these say so have rendered John and the Young Gentleman almost wild with anxiety — they talk of bringing Mothers bed of Eider down — Bears and White Wolf skins for matts and eggs in salt with thousands of other curious things as if they had these already in their possession.

The party set sail on the 6th of June and rounded the western shores of Nova Scotia, headed east past the big fort at Halifax on its southern coast, before turning northwards through the Canso Straits to set course for the Magdalen Islands and, just beyond, the famous gannetry at Bird Rocks. In another letter to his wife, written on the eve of his departure, Audubon had said, 'Thou may be assured that I will try hard to procure furrs for thy dear sister, Mary, and thyself — indeed I intend to try my hand at speculation in feathers and furrs if these can be procured with advantage hereafter.' A rising population in the United States ensured that, even if a life of affluence was not open to all, there was a ready market for affordable luxuries, including bedding and pillows made from the soft white feathers of sea birds. Fortunately perhaps for Audubon's reputation as the father of the movement for bird protection in the United States,[1] history does not relate whether he acted on his intention.

As they neared Bird Rocks on the 14th of June a violent squall suddenly descended on the party. While this did not dampen the enthusiasm of John Jnr. and his friend to effect a landing, Audubon himself was content to remain on the schooner, the *Ripley*, observing the Gannets from a distance and discussing their breeding habits with those on board who knew them well. These breeding grounds had been visited three centuries earlier, in 1534, by the French navigator Jacques Cartier (1494–*c.*1557; his name sometimes spelt 'Carthier') during his search for the elusive — and for all practical purposes illusory — North West Passage. Cartier related how:

> As we were about to hoist sail the wind turned into the north west wherefore we went south east about fifteen leagues, and came to three islands, two of which are as steep and upright as any wall, so that it was not possible to climb them: and between them there is a little rock.[2]

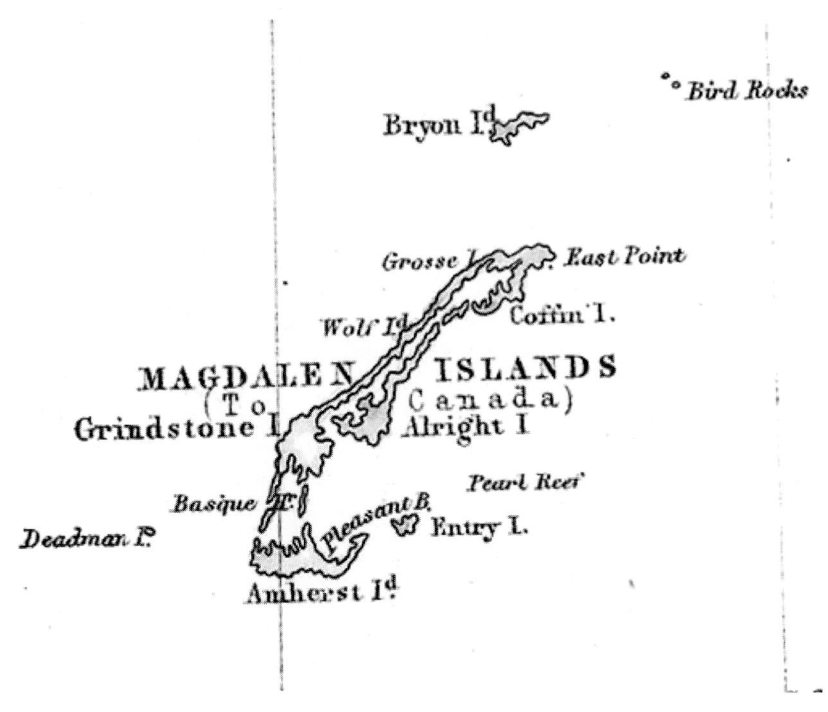

Fig. 1 The Magdalen Islands and Bird Rocks, from Keith Johnston's *General Atlas* (*c.*1860).

> These islands were as full of birds as any field or meadow is of grass, which there do make their nests, and in the greatest of them there was a great and infinite number of those that we call *Margaulx* [gannets], that are white, and bigger than any geese, which were severed in one part.

The sight which met Audubon's eyes complied fully with Cartier's description. 'No man,' he wrote in his journal later, 'who has not seen what we have seen this day can form the least idea of the impression the sight made on our minds.' Audubon learned that parties of fishermen were in the habit of landing annually to secure supplies of the adult birds to use as bait when fishing for cod. As a preliminary they would create a hullaballoo which so frightened the birds that they all took wing together, but many of them, finding that they had insufficient airspace to fly properly, descended to the water from which they could only take off with difficulty, even when the weather was calm. Then each man, armed with a club, would kill at the rate of ninety or a hundred birds an hour until either everyone was exhausted or they had a sufficient number to last them until the flesh became too putrid to be of any use. Because the topmost parts of the Bird Rocks stacks were almost inaccessible, and because the men had some reason to call off their slaughter after a time, annual visits of this kind did not prevent numbers of these birds returning to breed successfully.

Fig. 2 Oil painting of Audubon in 1833, aged 48, by Henry Inman.

Continuing northwards, Audubon's party proceeded to the large island of Anticosti and,[3] eleven days after leaving Eastport, they anchored at the Esquimaux Islands on the coast of Labrador. They remained six weeks in the region, finding the terrain difficult and fatiguing, the numbers of birds rather sparse, and the mosquitoes a perpetual annoyance. As ever, Audubon elicited what information he could from the local people about the Great Auk and, we can tell by his tone, he found what he obtained to be both interesting and surprising:

> When I was in Labrador many of the fishermen assured me that the 'Penguin', as they named this bird, builds on a low rocky island to the south east of Newfoundland where they destroy great numbers of the young for bait; but as this intelligence came to me when the season was too far advanced, I had no opportunity of ascertaining its accuracy.

In truth it had been a matter of mere chance that Audubon had been one of the few American ornithologists to realize that the Great Auk was a visitor to those coasts at all. As a result he had not been baffled, as certain European ornithologists were to be subsequently, by talk of 'penguins'. In 1830, during one of his visits to England, Audubon had learned from Henry Havell, brother of his engraver, Robert Havell, about a Great Auk which 'in extremely boisterous weather' he had hooked on the Newfoundland fishing banks during his crossing to England a year or so previously. The bird had duly been hauled on board where 'it was left at liberty on the deck. It walked very awkwardly, often tumbling over; bit everyone within reach of its powerful bill, and refused food of all kinds. After continuing several days on board, it was restored to its proper element.'

Realizing that the Great Auk would have to figure in *The Birds of America*, Audubon had acquired a specimen from a London dealer on which to base his illustration. This bird, one of a small spate of such examples to come on to the market in the early 1830s (after the volcanic eruption which destroyed the Geirfuglasker) undoubtedly came from Iceland. For many years, until American museums began buying up European collections, this specimen was one of only two such exhibits in the United States.

As the central and western parts of the country were opened up in the late eighteenth and early nineteenth centuries, it was natural for American ornithologists to turn their attention to the new species of land birds as they were discovered. Consequently, in the early days, the Great Auk seems to have been overlooked, in spite of being a regular winter visitor to the New England coast, and it was a long time before anyone realized that it had once been abundant as a breeding bird further north.

While near the mouth of the Matasquan River, in present-day Quebec province, Audubon observed certain fowling operations which he found deeply disturbing. From the *Ripley* he sighted some rocky islets thronged with Black

Fig. 3 A watercolour by Audubon of the Great Auk.

Guillemots (or 'Greenland Doves' as they used to be called on account of their pigeon-like winter plumage). He promptly made preparations to reach the rocks in order to ascertain how many eggs this species laid — a question of some controversy at the time — but was prevented for some hours by the force of the breakers. When eventually he was able to land, he found that a fowling party had got there before him and had entirely denuded the crevices of both eggs and adult birds. 'This war of extermination,' he wrote, 'cannot last many years more. The eggers themselves will be the first to repent the entire disappearance of the myriads of birds that made the coast...their summer residence.'

Audubon's party reached the Straits of Belle Isle, which separate Labrador from the long western peninsula of Newfoundland, on 20th August and found the first signs of approaching winter, in the form of icebergs. Doubling back, they hurried southwards down Newfoundland's western coast and anchored at St George's Bay where they spent a week and 'ransacked the country' for birds (as Audubon put it in a letter to his brother, Victor, the following month). He was impressed by the landscape which he found 'still more elevated, rugged and wild-looking' than the Labrador coast.

In the *American Gazetteer* (1762) may be found the following brief description of the topography of Newfoundland:

> The island is full of hills and mountains covered with pines so that the country can be traversed only in those parts where the inhabitants have cut a road through the woods. The trees of this species of pine seldom exceed 18 or 20 feet in height except those growing in the valleys, where they are sheltered from the piercing winds, which often are 40 foot high.

In spite of being on the same latitude as the English Channel, the climate in Newfoundland, in the absence of the Gulf Stream's warmth, is quite different from that of the British Isles: 'The cold during the winter is excessive here,' continued the account in the *American Gazetteer*, 'and the frosts, which are remarkably severe, set in about the middle of November and soon after the harbours and bays are entirely frozen.'

While in the west of the island Audubon had verbal corroboration of what the Labrador fishermen had told him regarding the breeding site of the 'penguins'. The information which Audubon gleaned only received wide circulation in Britain with the publication in December 1842 of the third volume of William Yarrell's *A History of British Birds*. Immediately those who sought specimens to enhance their collections commenced scouring their atlases to discover the whereabouts of 'the low rocky island' south east of Newfoundland.[4] Only the most alert of these ornithologists puzzling over the charts would, however, have noted the significance of a final sentence quoted by Yarrell from Audubon's account: 'An old gunner residing on Chelsea beach, near Boston, told me that he well remembered the time when the Penguins were plentiful about Nahant[5] and some other islands in the bay.'

With the introduction of the single word 'remembered' a disturbing element was introduced into what was the only remotely accurate account of the Great Auk to be given by an American naturalist of Audubon's era. It is, however, doubtful if Audubon himself had anything but the dimmest perception of the former abundance of the species which had once wintered in numbers as far south as Cape Cod.[6] In his brief *Synopsis of the Birds of America* (1839), he stated simply: 'Rare and accidental on the banks of Newfoundland; said to breed on a rock on that island' — a remark that is, in fact, contradictory because breeding birds, however rare, cannot be said to be 'accidental'.

The extent to which North American ornithologists remained ignorant about the particulars of the Great Auk's extinction is clearly shown in an article on the species by Professor James Orton in *The American Naturalist* in 1869. His readers were told that no one knew what it was, whether 'subsidence of strata, encroachments of other animals or climatal revolutions that caused the Great Auk to become extinct. We know of no changes on our northern coasts sufficient to effect the conditions necessary to the existence of this oceanic bird.' He then stated categorically: 'It has not been hunted down like the Dodo and the Dinornis [or Moa]' and concluded, 'we can only say with Buffon, it died out because time fought against it'. Twenty years later, an assistant curator at the United States National Museum, F.A. Lucas, who had visited the Great Auk's former breeding ground on Funk Island on behalf of the celebrated Professor Spencer F. Baird of the Smithsonian Institution, caustically remarked that there was but one accurate comment in the whole of Professor Orton's paper — that the bird, as was widely believed, was indeed extinct.

A century previously the sight of Great Auks off Newfoundland was not especially remarkable. When the British naturalist, Sir Joseph Banks, visited the territory in 1766, primarily to research its botany, he noted in his diary for the 7th of May that year, simply, 'birds that the sailors call Penguins'[7] near his ship HMS *Niger*. Further north, the penguin was a less frequent visitor, but Banks obtained a specimen in York Harbour (also known as Chateaux Bay) on the west side of the Straits of Belle Isle, Labrador, in August of the same year. Five years later, a local man, Captain George Cartwright, who for many years kept a journal of his extensive trapping operations in Labrador, wrote:

> *Monday, August 5th 1771*: During a calm in the afternoon, Shuglawina went off in his kyack in pursuit of a penguin; he presently came within a proper distance of the bird and struck his dart into it; but, as the weapon did not enter a mortal part, the penguin swam and dived so well that he would have lost both the bird and the dart had he not driven it near enough the vessel for me to shoot it.

This very matter-of-fact diary entry was written, as we shall see, by one of the handful of people who subsequently tried to save the Great Auk from extinction.

Although Newfoundland had the distinction of being Great Britain's first colony, it was viewed in most quarters as little more than a fishing station. However, as the eighteenth century had been an eventful one, a number of writers endeavoured to heighten public awareness in the mother country of its strategic importance and distinctive history. The first of these was Lewis-Amadeus Anspach, a Protestant missionary to the island, who privately published his *History of the Island of Newfoundland* in 1819. Anspach was a fine writer with an eye for detail which renders his accounts of the power struggles between Great Britain and France exceptionally vivid. It is regrettable, however, that he was only an indifferent natural historian:

> There was formerly on this coast a species of birds of the diving genus, which, from their inability to fly were always observed within the space between the land and the Great Bank and were once so abundant as to have given their name to several islands upon that coast, but they are now utterly extinct. They were known by the name of *penguins*...Captain Cook found these birds in great numbers near Terra [*sic*] del Fuego; his people gave them the name of 'race-horses' on account of the great swiftness with which they were observed to run upon the water.

Anspach's supposition that the penguin of Newfoundland and the penguins of the Southern Hemisphere were identical would undoubtedly have suggested to most of his readers that what had occurred was merely a local extinction and that the species was plentifully represented elsewhere. Unfortunately, there were too few naturalists with the relevant experience to notice that between the time of Sir Joseph Banks' visit and the end of the wars against revolutionary France, either a natural cataclysm had occurred or some dreadful work of destruction had been carried out, so as to effectively annihilate a whole species, not simply a local population. An authority on the Great Auk in the later nineteenth century, Professor Alfred Newton of Cambridge University, pertinently remarked: 'The exterminating process is generally one that attracts little or no attention until the doom of the victim is sealed.'

As Anspach's work was published a decade before Henry Havell caught the Great Auk which he so graphically described to Audubon, clearly the species was not 'utterly' extinct. There can, however, have been precious few[8] in existence at that time, either on the fishing grounds of Newfoundland or among those rocky skerries in northern Europe that the Great Auk had colonized.

Notes

1 Advocates of bird protection in the nineteenth century were known by their detractors as 'Audubonites'.

2 It was on this rock that Audubon's son and his friend landed on the 14th of June.

3 At nearby Blanc Sablon, John Woodhouse Audubon found, situated in low tangled fir bushes, some disused nests of the Labrador Duck, a beautifully marked species which became extinct suddenly and unexpectedly in the 1870s. Almost certainly the excessive killing in the eighteenth century of this and other species of wildfowl during the autumn moult, in order to obtain their feathers, was a contributory factor.

4 The principal breeding ground of the Great Auk in Newfoundland's territorial waters, Funk Island, in fact lies to the *north* east off Cape Freels; as Audubon learned about this locality when in Labrador, it is possible that the words spoken were 'to the south east *off* Newfoundland', a phrase which would have been entirely accurate.

5 Nahant Island, Massachusetts Bay.

6 The seventeenth-century naturalist and illustrator, Mark Catesby, noted that the Penguin was an irregular winter visitor as far south as the coast of Carolina.

7 This is an echo, conscious or unconscious, of a phrase found in the account of the Great Auk in John Ray's *The Ornithology of Francis Willughby* (1678).

8 There is uncorroborated evidence to suggest that the Great Auk was still common, locally, off Fogo Island, on the north east coast of Newfoundland, as late as the 1820s. The American ornithologist, George A. Boardman, was told by the Revd William Wilson (who had been a Methodist missionary in the area from 1818–23) that he 'saw the Penguin during the whole of his stay in the island in considerable numbers' and that it 'was quite common for the boys to keep them tied by the legs as pets'. Unfortunately, the description Mr Wilson gave of the bird — '...a coal-black head and back, a white belly, and a milk-white spot under the right eye' — seems based more on the earlier writing of the author of *The English Pilot* (1728 and subsequent editions) than it does on personal observation. Consequently, whether the birds seen during his stay were Penguins or some other species of Auk must remain an open question.

CHAPTER 4

Westward ho!

One of the first popular writers on the history of discovery in North America was John Reinhold Forster, a naturalized Briton of German birth. Resettled in London after serving as ship's naturalist during Captain Cook's second Pacific voyage, Forster wrote lucid accounts of the sixteenth-century venturers to the New World. His *History of the Voyages and Discoveries Made in the North* (1786) was derived from the accounts compiled by Richard Hakluyt, the historian and 'apostle of English expansion', which had appeared nearly two centuries earlier.

Referring to the Great Auk, Forster informed his readers:

> [W]e also find it mentioned that a person by the name of Hore set sail in 1536 from London with two ships the *Trinity* and the *Minion* about the latter end of April. They arrived at Cape Briton [i.e. Cape Breton Island] and from thence went to the north-eastward until they came to Penguin Island...which was named thus after a certain kind of seafowl, which the Spaniards and Portuguese called Penguins on account of their being so very fat, and which used to build their nests and to live in astonishing quantities on this little rock.

The explanation of the bird's name given by Forster is no more than a restatement of the opinion, seldom if ever challenged, of seventeenth-century authorities such as Carolus Clusius and Francis Willughby. This traditional explanation was, however, challenged by Lewis-Amadeus Anspach in his laconic account of the extirpation of the Newfoundland penguins with which we concluded the previous chapter. Anspach drew attention to the opinion of some writers who asserted that the name was derived 'from the Welsh in which language it signifies white head'. It is only possible, Anspach argued, to make sense of the traditional explanation of the name if it is derived from the Latin 'pinguis'. Implying that it was most unlikely that the first generation of seafarers to venture into the waters off Newfoundland would have given Latin names to the unfamiliar creatures which they encountered, Anspach remarked that it was more natural to suppose that the Spaniards and Portuguese would have made use of the vernacular word 'gordo' which is common to both languages.

Foremost among those writers advocating a Welsh origin for 'penguin' was the sixteenth-century supporter of westward expansion, Sir George Peckham. Hoping that colonization of America would alleviate the suffering of his fellow Catholics under Elizabeth I, Peckham compiled a *True Reporte* of the annexation of Newfoundland by Sir Humphrey Gilbert in 1583 as a consequence of which the island became England's first transatlantic dominion. As early as 1566 Gilbert had written a discourse on the possibilities of opening up trade with the east by a northerly route. He believed implicitly in the existence of a passage to the fabled wealth of 'Cathay' (China) 'by the Northwest from us through a sea which lieth on the north side of Labrador'.

We can be certain that all those who went in search of this route in the sixteenth century timed their voyages as far as possible to coincide with the breeding season of the Great Auk on the 'little rock' mentioned by Forster, supposing this to be the same as Audubon's 'low rocky island'. This granite outcrop north east of Newfoundland was strategically placed to facilitate such exploration. Finding this supposed north westwards route, claiming the northerly latitudes of North America for the English crown, and discovering mineral wealth were Gilbert's three ultimate ambitions. There was, too, a further service he could render his native land:

> We might inhabit some part of those countries and settle there such needy people of our country which now trouble the commonwealth, and through want here at home are inforced to commit outrageous offences, whereby they are daily consumed with the gallows.

The use of newly acquired territory overseas as a means of getting rid of 'surplus population' was typical of all the European powers in the sixteenth century and, indeed, throughout the age of imperialism. Inevitably, the first colonists were not always ideally suited to the task and throughout the centuries the crudest forms of exploitation and violence were employed by people who might well have been considered as the scum of society in the land of their birth. If this tended to have a negative impact on the native wildlife as well as on the native peoples of those newly won territories, then this must be seen as a direct and predictable consequence of the government policy of the day.

Leaving the manuscript of his discourse unpublished for a decade, Gilbert served almost continuously in Ireland. There he rose to the rank of colonel and obtained his knighthood for his ruthlessness towards the rebellious inhabitants of that unhappy isle in support of a policy intended, in the words of the then Lord Deputy of Ireland, to make the very word Englishman 'more terrible now to them than the sight of an hundred was before'. We can well imagine that, seeking eventual relief from this brutalizing existence, Sir Humphrey Gilbert longed for the open sea and the possibility of finding wealth and fame in the New World.

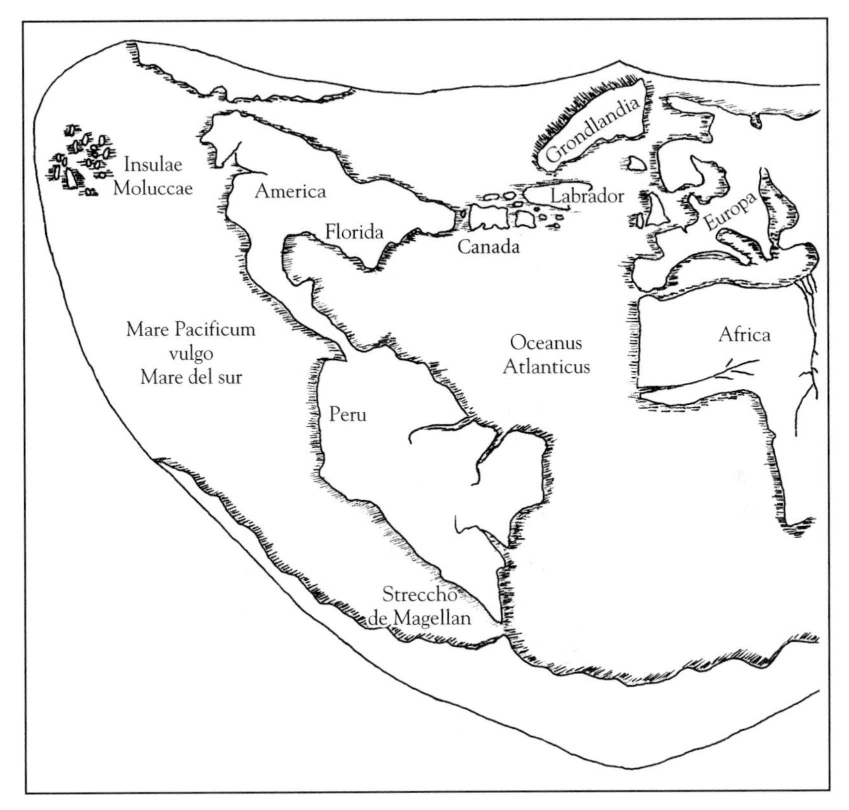

Fig. 1 Map of Europe, Africa, and the New World showing the supposed North West Passage to the Moluccan Spice Islands. From Sir Humphrey Gilbert's *Discourse of a Discovery for a New Passage* (1566).

Although it is probable that Spanish fishermen and Basque whalers frequented the Newfoundland fishing banks as early as the middle of the fifteenth century, the official discovery of Newfoundland was made in 1497 by John Cabot (Giovanni Caboto) at the behest of the then King of England, Henry VII. Expecting a rich booty in the form of precious metals the king had been disappointed to receive news of nothing but an abundance of cod and consequently had rewarded his admiral with a meagre pension. By the time that Sir Humphrey Gilbert voyaged to Newfoundland 86 years later, Queen Elizabeth still desired wealth to rival that of Spain, hoping to 'plant Christian inhabitants in place convenient upon those large and ample countries extended northwards from the cape of Florida...not in the actual possession of any Christian prince'.

The Queen's ambition evidently extended beyond merely annexing the territory discovered by Cabot whose adjacent waters had become an international fishing ground. A detailed account of the voyage from the pen of one of Gilbert's

Fig. 2 Sir Humphrey Gilbert who observed breeding Great Auks at the end of July 1583. Note the motto 'Quid non' — 'Why not?'

officers, Edward Hayes, captain of the *Golden Hind*, includes a description of their first sight of Great Auks:

> We had sight there of an island named *Penguin*, of a fowl there breeding in abundance, almost incredible, which cannot fly, their wings not able to carry their body, being very large (not much less than a goose) and exceeding fat: which the French men use [*sic*] to take without difficulty upon that island, and to barrel them up with salt. But for lingering of time we had made us there the like provision.

Subsequently Gilbert's flagship, the *Delight*, foundered in thick fog on one of the many shoals on that coast. Salvaging what he could, but not the valuable specimens his German mineralogist claimed to have found, Gilbert transferred to the *Squirrel*, a support ship of less than ten tons and insisted, despite the misgivings of Edward Hayes and others, on recrossing the Atlantic in that trusty little frigate rather than transferring to Hayes' ship. As the party neared the Azores they encountered 'very foul weather and terrible seas'. Gilbert was observed seated calmly, reading, in the after part of the vessel and as the billows threatened to engulf them was heard to encourage his men with the words, 'Cheer up lads! We are as close to Heaven by sea as we are by land'. That night the *Squirrel* went down, disappearing without trace.

Fig. 3 An English warship, *c.*1580.

Disconsolate at the loss of their commander-in-chief, the crew of the *Golden Hind* continued their homeward voyage to be reunited at last with those of another ship of the original squadron, the *Swallow*, which had been sent home early with the sick. Hayes and the other officers ensured that a full report was made to Gilbert's half-brother, Sir Walter Raleigh.

Gilbert's voyage might easily have been regarded as a *casus belli* by an increasingly hostile Spain. In 1497, at a time when Henry VII had conducted a pro-Spanish foreign policy, Madrid's ambassador in London, on hearing of Cabot's voyage, had demanded an audience with the King in order to remonstrate. Whatever his goal, John Cabot was clearly acting in violation of the revised bull of Pope Alexander VI of 7 June 1494 which, in order to minimize the potential for conflict, had divided the world into two hemispheres, each the exclusive preserve of one of the only two maritime powers at that date, Portugal and Spain. All lands discovered thus far or to be discovered in the future were deemed to belong to one or other of these two nations. When Henry received the Spanish ambassador he could do nothing but feign an apology, Cabot having set sail for the New World already.

By the late 1570s, when Gilbert received his commission from the Queen, remonstrations from Spain would have been coldly rebuffed. Styling herself the Supreme Governor of the Church of England, Elizabeth did not feel bound by Papal bulls, past or present. In addition, she shared the view held by many in Europe, Catholic and Protestant alike, that the high seas should be in principle — as they were in practice — open to mariners of all nations. If Gilbert's voyage was regarded as a provocation by a jealous Spain, it was one which it was expedient to pass over for the time being since a war between the two nations was increasingly seen as inevitable. Strange as it may seem, the Great Auk was destined to play a small but not insignificant role in this conflict, an unwitting pawn in the deadly international rivalry of that era.

Like other Tudor monarchs before her, Elizabeth sought to lend a veneer of legal respectability to dubious actions taken in furtherance of state policy. In order to justify English expansion into the New World an argument almost breathtaking in its audacity was devised by the Queen's apologists: the discovery of America, it was said, had been made by a Briton 'lineally descended from the blood royal', Madoc ap Owen Gwyneth, more than three centuries before Columbus. 'For no other nation,' wrote Sir George Peckham in his *True Reporte* of Gilbert's voyage, 'can truly by any chronicles they can find make prescription of time for themselves before the time of this prince Madoc'.

The illegitimate son of a twelfth-century prince of north Wales, Madoc is reputed to have landed at Mobile Bay, in present day Alabama, where he left a number of companions under his brother Rhiryd while he returned for more volunteers. No more is known of Madoc and the fate of his party has been the subject of much conjecture, some suggesting that they formed a Welsh-speaking tribe of Indians, the Mandans, who finally perished in a smallpox epidemic as recently as 1838. Sir George Peckham, however, had no doubt that Madoc was no less a person than the legendary ancestor of the Aztecs. In support of this contention Peckham cited the opening remarks of a speech made by Monteczuma, the last Aztec emperor, when he yielded his dominion to the *conquistador* Hernan Cortes. Addressing his subjects he said:

> You ought to have it in remembrance that we are not naturally of this country, nor yet is our kingdom durable, because our forefathers came from a far country and their king and captain who brought them hither returned again to his natural country saying that he would send such as would rule and govern us if he by chance returned not.

According to Peckham, Monteczuma's words effectually corroborated the historical basis of Madoc's voyage which seemed evident from the occurrence of supposedly Welsh names in North America. These include 'gwyneths' (a type of fruit) and, significantly, the Island of Pengwyn. Concerning the latter, Peckham decided to elaborate: 'There is likewise a fowl in the said countries called at the same name at this day, and is as much to say in English as 'whitehead' and in truth the said fowls have white heads.'

However ingenious his *True Reporte*, Peckham was merely a disseminator of these theories, not their author. The distinction of having been so belongs to Dr John Dee, a gifted mathematician and geographer as well as a dabbler in sorcery and spiritualism. Retained at court as a consultant on state affairs, Dee drew up for the Queen a series of 'titles' in which he traced Elizabeth's right to a number of territories on the grounds of modern cosmographical calculation and Celtic tradition. In this way Scandinavia, the entire Arctic, and the islands westwards to what he called 'Estotiland' were claimed for England as having been conquests of none other than the 'once and future king' of British legend, Arthur. Dee's *Title Royal* of 1580 claimed all the newly discovered territories in North America on the slender evidence of Madoc's voyage.

Such evidence is slender indeed: the Flemish author of a popular thirteenth-century satire *Reynard the Fox*, known to posterity as Willem the Minstrel, described himself in the prologue to *Reynard* as 'Willem the author of Madoc'. A fragment of what appears to be a synopsis of this ballad in French was discovered at Poitiers in the seventeenth century. We learn that Madoc discovered an island paradise of music and love and returned subsequently to recruit more volunteers. Of topographical details which relate to North America there are none. However, 'the warm sea in which plants do grow' (possibly a reference to the Sargasso Sea), when taken together with the fact that the nails of Madoc's ship are described as having been made of stag horn in order to minimize compass variation, does suggest that exploratory voyages westward were made by intrepid souls in the centuries before Columbus.

Certainly Dee had presented his *Title Royal* at an opportune time. Falling into disrepute later in the sixteenth century for publicizing his supposed communications with spirits, Dee was to be accorded rather less patronage thereafter. However he was never completely disgraced — he had served his royal mistress well.

John Reinhold Forster's explanation of the derivation of the word 'penguin'— the birds being so called because they were 'so very fat'— finds acceptance in Spain and Portugal to this day. Besides the word 'gordo', to which Anspach alluded, there exists another, 'pingüe', which is clearly derived from the Latin 'pinguis' and common to both languages. The Portuguese phrase 'lucros pingües' ('fat profits') shows clearly how the word is used. It would seem that for fishermen from Portugal and Spain, 'el Pingüino'(as they called the Great Auk) suggested a satisfying meal after weeks of nothing but ship's rations.

In view of the tendentious nature of Sir George Peckham's arguments, it might seem appropriate to dismiss the notion of a Welsh derivation for 'penguin' as groundless. To do so however would be to overlook the fact that the Breton name for the Great Auk's smaller congener, the Razorbill, is 'Pen-gwenn', a term which translates in exactly the same way as the Welsh 'Pen-gwyn'. It seems that on adopting the name the fishermen from Brittany preferred the Celtic to the Iberian derivation. This may be no more than an instance of solidarity between different peoples of the Celtic fringe of Northern Europe.

There is, however, some merit in the theory of a Welsh — or possibly Cornish — origin for the name. It is not inconceivable, in view of the variety of Celtic and Iberian tongues spoken in the region of Newfoundland in the sixteenth century, that similar sounding words in different languages were assumed to have similar meanings. Perhaps in 'pingüe' and in the Welsh 'pengwai' we have two such linguistic 'false friends'. On account of its sharply hooked beak the Razorbill is known in Welsh as the 'gwalch y pengwai' — 'gwalch' signifying a Hawk and 'pengwai', the herring. As the Great Auk was similarly endowed with a decurved upper mandible, the name might have been misapplied by anyone of Welsh origin who joined the annual fishing expeditions which set sail from south west England. In the course of time, the term might have been taken up by the Spanish and Portuguese, believing the name to signify the same degree of plumpness as 'pingüe'.

Given the much greater number of Iberians in those waters in the early days, a contrary process is more likely to have operated, 'pingüino' suggesting 'pengwyn' or one of its variants in the other southern Celtic languages, Cornish and Breton. While this remains a matter of conjecture, one fact remains certain — Englishmen were pronouncing and spelling the word as '*pen*guin', as distinct from '*pin*guin', before Sir George Peckham wrote his *True Reporte* in 1583. Whatever self-serving arguments may have been employed by Englishmen thereafter, these alone do not explain the distinctly Celtic flavour of the traditional Newfoundland name for the Great Auk.

CHAPTER 5

A visit to Funk Island

In 1839, a young Norwegian naturalist, Peter Stuvitz, was commissioned by his government to conduct research into the Newfoundland fisheries which, it was felt at the time, were posing a significant threat to the home fishing industry. Safely arrived in St John's, Newfoundland's small capital, Stuvitz did not limit himself to purely economic matters but, keeping a detailed journal on every branch of natural history and working always from firsthand observation, proved himself a potentially great scientist. Sadly, however, about a year after his arrival he contracted tuberculosis, an illness from which he never recovered. His death, which occurred on 21 June 1842, was probably hastened by his exposure to the worst of the elements which he had been prepared to endure in the pursuit of his researches. In the meantime he had circumnavigated Newfoundland's extensive coastline, collected (like Audubon) on the Labrador coast, and had conducted research in the area of most concern to those at home in Norway — the Great Bank. This was described in *The American Gazetteer* (1762) as:

> ...a vast mountain under water, about 530 miles in length and 270 in breadth. The depth of the water is very unequal, from 15 to 60 fathoms. The bottom is covered with a vast quantity of shells and frequented by large shoals of small fish, most of which serve as food to the cod which are here in amazing plenty; for though two or three hundred vessels have been annually loaded with them, during the last and present centuries, yet the prodigious consumption has not lessened their plenty. We cannot help observing, that this fishery is a mine of greater value than any of those in Mexico or Peru.[1]

When consulting maps of Newfoundland in preparation for his field trips, Stuvitz was baffled and intrigued to find the name 'Penguin Is.' inscribed at different points off the coast. Later he wrote:

> At St John's they told me that the Penguin had actually remained near the coast and, moreover, that Funks Island was one of the islands where they could be met with in greater numbers, that once upon a time they pursued it every year for its down and feathers and that there was still on the island some mounds of the bones and skeletons of this bird.

> Briefly adding together, in a general way, the maps and traditions respecting matters of concern to Ornithology, I considered that the name 'Penguin' could have been given to a competely different bird from the one which we assign to this species.

Evidently Stuvitz communicated what he had been told to staff at Christiania (i.e. Oslo) University who replied that the birds were certainly not Penguins properly so-called, as these were known to occur only in the Southern Hemisphere, as had been finally established by the French naturalist Baron Georges Cuvier (1769–1832). Obviously still puzzled by the references, and possibly feeling that his academic reputation was at stake, Stuvitz decided that he must visit Funk Island — or 'The Funks' as the principal island and two associated islets were often called — to obtain evidence in the form of bones for those at the University who were overseeing his researches. Selecting a suitable opportunity with regard to the weather, he set out from St John's on 30 June 1841 and arrived at his destination at noon the following day. The birds breeding at the beginning of July[2] would have comprised abundant numbers of Arctic Terns and, in the crevices, Leach's Storm-Petrels, a few Kittiwakes on the low cliffs, together with numbers of different kinds of Auk (Puffins, Razorbills, both Common and Brünnich's Guillemots, and, according to Stuvitz's journal, Black Guillemots, a species which no longer breeds on the island today).

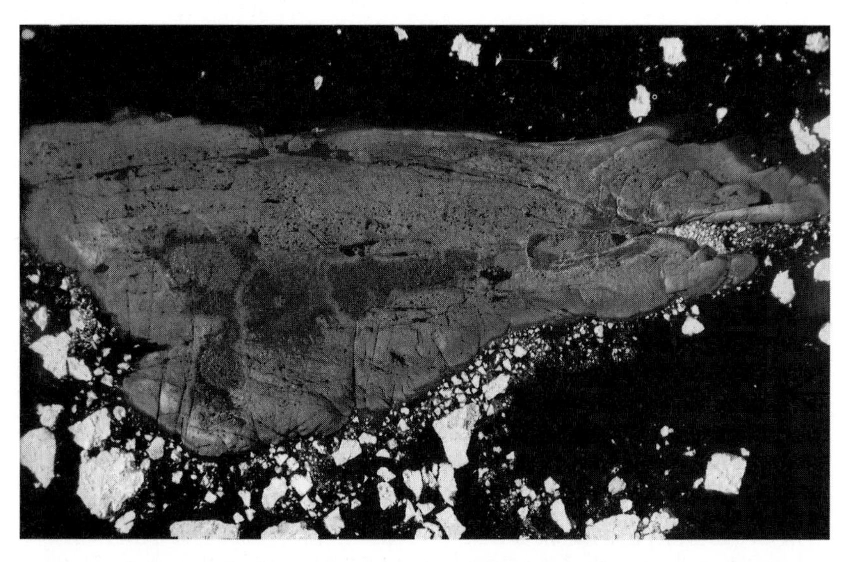

Fig. 1 Aerial photograph of Funk Island, Newfoundland, 19 June 1972, surrounded by spring sea-ice. Great Auks may once have concentrated for breeding along the median line from West to East now occupied by breeding Common Murres (Common Guillemots). (Photo by D.N. Nettleship)

Funk Island's latitude (49° 45′ north) had been pinpointed, almost exactly, in 1535 — nearly half a century before Sir Humphrey Gilbert's arrival in Newfoundland — by the Breton navigator Jacques Cartier. In 1541, at the request of the King of France he had made a third voyage, this time of a military character, in search of the North West Passage and the access to the Spice Islands which it was hoped it would give.[3] Separated from their flotilla by bad weather, those in the general's ship reached Newfoundland on the 7th of July and, in Cartier's own words:

> ...came to the island called the Island of Birds, which lieth from the mainland 14 leagues.[4] This island is so full of birds that all our ships might easily have been freighted with them, and yet for the great number that there is, it would not seem that any were taken away. We to victual ourselves filled two boats. This island hath the pole elevated 49° and 40′.

Evidently Cartier gave the latitude of the 'Isle des Oyseaux' so that future venturers would know where to obtain a ready supply of fresh meat. Funk Island is marked as 'Penguin Island' on maps by John Dee of *c.*1580 and on Michael Lok's map of the New World (1582) simply as 'Aves' (Birds).

On approaching to within a dozen miles — the furthest distance from which Funk Island can be observed, even on a fine day — the island in 1535 would have appeared much as it did to Stuvitz in 1841 and, indeed, as it does today. A flat sliver of pinkish granite, half a mile long by a quarter of a mile wide, its main axis lying south west to north east, Funk Island viewed from a distance has been likened to a smooth-bottomed, upturned saucer. The cliffs on the south side of the island reach no more than 50 feet in height and are only precipitous at the east end. These cliffs, however, continue vertically down another 150 feet below the surface of the sea — a feature which ensures that there are no breakers, except in extreme weather conditions, but instead at all times a strong swell. Although there is a gentle gradient on parts of the southern shore, the principal areas of shelving rock are on the northern and southern sides of the island. The gradient being sufficiently gentle, it was to these points that, for countless generations, the innumerable hosts of Great Auks came ashore to breed.

The oldest reference to the Great Auk on Funk Island is from Cartier's narrative of his first voyage in 1534. In this account three kinds of bird were singled out as being worthy of special reference. The first, 'Margaulx' ('which bite even as dogs'), were, as we have seen, Gannets. The identity of the second species is not quite so straightforward, but the birds he named 'Godertz' (in English usually spelt 'Godets') which 'gather themselves together on the island and put themselves under the wings of other birds that are greater' were one (or perhaps both) of the smaller species of Auk, for Guillemots, otherwise known as Murres, do tend to nest in ones and twos among Gannets, as well as in numbers on their own. (In French-speaking Canada, Razorbills are still referred to as 'Godes'.) A third kind which were called 'Apponath'[5] by the Beothuk Red Indians of Newfoundland, Cartier described in the following terms:

> ...as big as *Pies*,[6] black and white, having the beak of a Raven (ayans le bec de Corbeau): they lie always upon the sea; they cannot fly up high because their wings are so little and no bigger than half ones hand, yet do they fly as swiftly as any birds of the air level[7] to the water (à fleur d'eau).

Cartier added that these Apponats were 'exceedingly fat' and that: 'in less than half an hour we filled two boats full of them as if they had been with stones: so that besides them that we did eat fresh, every ship did powder and salt four or five barrels full of them'.

The abundance of sea birds on this small island made a lasting impression on Jacques Cartier. It was still relatively early in the year — the 21st of May — and the ice around the island was still in the process of breaking up, yet the birds were there in such plenty that:

> ...unless a man did see them, he would think it an incredible thing: for albeit the island...be so full of them that they seem to have been brought thither and sowed there for the nonce [i.e. expressly for that purpose], yet are there an hundred times as many around the island and in the air as there are on it.

As it seems unlikely that the Great Auks and other sea birds would have been breeding before the ice had receded from the 'Island of Birds', it may be that Cartier confused the details of his two visits to these waters in 1534 and 1535. Be this as it may, some time later Cartier must have related these details to his compatriot, André Thevet, who recalled them in *Les singularitez de la France antarctique* (see Chapter 1), when he came to describe the birds he termed 'Aponars', which he claimed were then to be found on Ascension Island:

> [O]n a time three great ships of France, going to Canada, did lade each of them two times their cockboats with these birds on the brink of the said island, and it was no mystery to go into the island and to drive them before them to their boats like sheep (ainsi que moutons à la boucherie).

Cartier's accounts of his voyages were not collated and published in France before the appearance of Marcus Lescarbot's *Histoire de la Nouvelle France* (1609). They had, however, already been published in English by Richard Hakluyt. Hakluyt's account of the voyage in 1536 of the two ships *Trinity* and *Minion* commanded by Robert Hore (a summary of which by John Reinhold Forster was given in the preceding chapter) was derived from the last survivor of that expedition, Thomas Butts. Such was Hakluyt's zeal to obtain information from participants in the events he described that in 1578 he rode the two hundred miles from London to Norfolk to conduct the interview personally. Butts told him that once the ships had reached Cape Breton Island:

> ...shaping their course thence north-eastwards they came to the island of Penguin, which is very full of rocks and stones, whereon they went and found it full of great fowls white and grey, as big as geese, and they saw infinite numbers of their eggs. They drove a great number of their fowls into their boats upon their sails and took up many of their eggs. The fowls they flayed...[and] dressed and ate them, and found them to be very good and nourishing meat.

In November 1578, quite coincidentally, Hakluyt received a letter from a regular visitor to those parts, Anthony Parkhurst, who had developed a scheme for fortifying Belle Isle near the Labrador coast, thereby making the English nation 'lords of the whole fishing in small time if it do so please the Queen's majesty'. The only obstacle to realizing such a plan was the much greater number of vessels from Portugal, Spain, and France than those from the west of England[8] so that a fivefold increase in the number of Englishmen in those waters was a necessary first step. Much of Parkhurst's description of the attractions of the country and its supposed fertility and clement climate is unashamedly propagandistic in tone and we should take with a pinch of salt his account of the ease with which the island's bird life could be exploited:

> There are seagulls, murres, ducks, wild geese, and many other kinds of birds store too long to write, especially at one island named *Penguin* where we may drive them on a plank into our ship as many as shall lade her...These birds are also called penguins, and cannot fly; there is more meat on one of these than in a goose: the French men that fish near the Grand Bay do bring small store of flesh with them, but victual themselves always with these birds.

In due course others proved receptive to these plans, for Parkhurst's letter played its part in persuading the Privy Council that it was expedient to claim sovereignty over territory in the New World. Although it took Sir Humphrey Gilbert another five years to set sail, it is significant that 1578 was also the year in which he received from Elizabeth I his letters patent which, for a period of six years, granted him extensive rights in North America in return for royalties paid to the crown.

Prior to the arrival of Europeans, the Great Auks on this island had not, of course, entirely escaped pursuit for the sake of their 'good and nourishing' meat. After describing the 'Apponatz', 'Godertz', and 'Margaulx' encountered on his voyage of 1534, Jacques Cartier observed, 'And albeit the said island be 14 leagues from the mainland, notwithstanding bears come to eat of the said birds'. Equally remarkable is an entry, dated 5 July 1785, in the *Journal* of Captain Cartwright (mentioned in Chapter 3):

> It is a very remarkable thing (yet a certain fact) that the Red, or Wild, Indians of Newfoundland should every year visit that island; for, it is not to be seen from the Fogo hills, they have no knowledge of the compass, nor even had any intercourse with any other nation, to be informed of its situation. How they came by their information, will most likely remain a secret among themselves.

Cartwright, alas, wrote truer than he knew for in little over a generation the Beothuk Indians were no more, having retreated before a technologically superior civilization and fallen victim to the illnesses, especially smallpox and tuberculosis, which Europeans unwittingly brought with them. The last known Beothuk Indian died in 1829.

As he neared Funk Island, Peter Stuvitz would have had to choose his landing place according to the direction of the wind. In a northerly blow a tiny cove on the south west side of the island can be used; alternatively, if the wind is coming up from the south a fissure on the north side, only a hundred metres or so from the promontory at the north eastern end of the island, and known to the Newfoundlanders as 'The Bench', leads, by means of a well-trodden path over a metre wide, from sea level up to the safety of the rocks. Having an interest in geology, Stuvitz would have noted that on closer inspection the pink granite of which the entire island is composed revealed a high density of black mica in small flecks, only rocks drifted on to the island by pack ice during the winter appearing at all different. Round his head, too close for comfort, the Arctic Terns would have hovered and swooped in defence of their eggs and everywhere on the island, away from the cliffs, facing windward, he would have been assailed by a slightly nauseating 'pigeon-house' smell as if in a vast aviary. Where the Guillemots bred on the south east side of the island, a stench reminiscent of ammonia might have seemed overpowering at times. It is hard to imagine how much stronger this reeking aroma would have been during the first decades

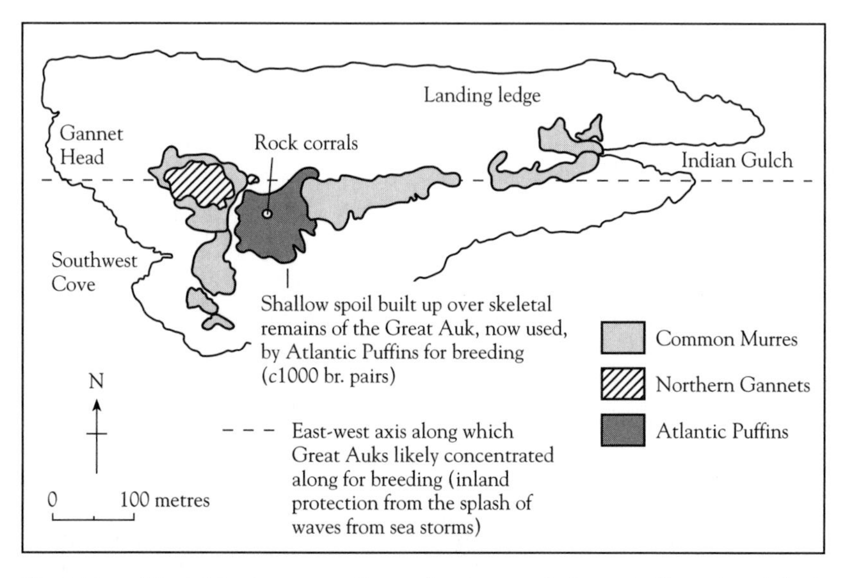

Fig. 2 Map of Funk Island corresponding to the photograph shown in Figure 1 — latitude 49° 45′ N, longitude 53° W. (From a sketch by D.N. Nettleship)

after European settlement of Newfoundland, when there would have been many more birds (including tens of thousands of Great Auks) than there were in 1841. The name of the island is, however, testimony enough, for a little-known meaning of the expressive word 'funk' is 'a strong, unpleasant smell'.

Looking across the island from any suitable vantage point, Stuvitz would have observed an undulating surface created by two fissures running roughly parallel south west to north east into which rainwater runs forming a few pools and some patches of boggy vegetation. It is possible, though by no means certain, that the more northerly of these two fissures would have created too great an obstacle for the Great Auks to surmount, thereby restricting the entire breeding population to the north western portion of the island. A few patches of tough grass and a flora hardy enough to withstand a windswept environment would have been observable in the sheltered spots, as would sandy accumulations composed of millions upon millions of tiny fragments of eggshell in the dry hollows. In the south western corner of the island lies Gannet Head, but Gannets had long since deserted this breeding site by the time of Stuvitz's visit, their nests having proved too easily accessible. Happily, there is today a thriving gannetry once more on the island, a few 'pioneering' birds having commenced occupation of this traditional site in the 1930s and the breeding population increasing rapidly thereafter.

On the ridge above Gannet Head is the only area of what could be termed soil on the entire island and it must have been with some sense of horror that Peter Stuvitz realized why it was that the plantain and a bit of blue-flowering weed grew more luxuriantly at this spot than elsewhere on the island. In his *Journal* he wrote that the reports he had received in St John's concerning the mysterious bird whose identity he hoped to settle:

> ...conformed to a certain extent to the truth, for I found on the west side of the island some remains of skeletons which, judging by their shape, had to have come from penguins. It is on the south west side that one finds a little vegetation and enough soil to produce a flora — a thoroughly wretched one in truth — but it is there that I discovered in great number the remains of bones and it is probably due to the destruction of the creatures to which they used to belong that the thin covering of vegetated soil that one encounters at this spot owes its origin.

Evidently Stuvitz was still mystified as to the identity of the species and sent a representative sample of bones to Christiania University. There they were promptly identified correctly as belonging to the Great Auk, but little importance was attached to this at the time for the species was regarded as common enough in the uninhabitable regions of the North even if it was on the verge of extinction in Europe itself. Consequently, the information that Peter Stuvitz had supplied was known for a number of years but to a handful of people. However, at about this time a number of Great Auk bones were also found in a prehistoric

Fig. 3 Remnants of the Great Auk corrals on Funk Island showing tussock grass vegetation that has built up over the skeletal remains of countless discarded Great Auks (18 July 1969). (Photo by D.N. Nettleship)

kitchen midden at Meilgaard in Jutland, northern Denmark. Academic staff at Christiania University were happy to oblige their colleagues at Copenhagen by sending some of Stuvitz's examples for comparison. These were received in 1844 and prompted a professor there, Japetus Steenstrup, to undertake extensive research with a view to establishing definitively the species' range. The fruits of his labour were not published until 1857 and so it was that many misconceptions on the part of the scientific community regarding the Great Auk's distribution only began to be dispelled after the last representatives of this species had disappeared from the face of the earth.

Steenstrup not only established that the most northerly breeding ground of the Great Auk did not even reach the southern boundary of the Arctic regions which had 'hitherto been assumed to be the proper home of this bird' but also that its breeding range within recent historical times probably consisted of its last strongholds only. Consequently, plotting a detailed map of where it once bred is simply not possible. Once the Great Auk had ceased to breed on some of the islands round Newfoundland, its association with the name 'penguin' was lost. Even where the name is retained to this day, such as for the group of islands off Cape de la Hune off south west Newfoundland, there is, hearsay excepted, an absence of any tradition, whether written[9] or oral, concerning the bird which must have been extirpated there long before the Newfoundlanders discontinued their annual expeditions to Funk Island.

Fig. 4 The father of modern Great Auk studies — Professor Japetus Steenstrup (1813–97). (Photograph by Reinhard Ruge)

For a reference to a breeding site for the Great Auk other than Funk Island, we must turn again to Lescarbot's account of Jacques Cartier's first voyage in 1534, during which this French navigator landed, as mentioned in Chapter 3, on the Bird or Gannet Rocks in the Gulf of St Lawrence. At that time there appear to have been not two, but three islands, one of them low-lying. Cartier recalled:

> Toward the shore there were of those Godetz and great Apponatz, like to those of that island that we above have mentioned [the Island of Oyseaux — Funk Island]: we went down to the lowest part of the least island where we killed about a thousand of those Godetz and Apponatz. We put into our boats so many of them as we pleased, for in less than one hour we might have filled thirty such boats of them.

Another account of the Great Auk on Bird Rocks comes from the pen of a Franciscan friar, Gabriel Sagard Théodat, who sailed close by in 1624 *en route* for Canada. He observed that there were:

> ...certain kinds there which cannot fly at all (ne peuvent presque voler), and which one can easily overpower by blows with clubs (à coups de bastons), as did the sailors of another ship, who before us had used their shallop for that purpose, and several barrelfuls of eggs, which they found in the nests; but they thought

> they would collapse with faintness (ils y pensèrent tomber de foiblesse) on account of the extreme stench of the dung of the said birds.

It is not difficult to see that as time went on this was increasingly the lot of the penguins to the west of Newfoundland and a series of local extinctions on its breeding grounds in those waters had probably occurred by the end of the seventeenth century or soon after.

It is likely that the Great Auk once bred also near the coast of Labrador for in the *Memoirs of North America* by Baron de Lahontan is the following curious reference:

> The Moyacks are a sort of fowl, as big as a goose, having a short neck, and a broad foot; and, which is very strange, their eggs are half as big again as a swan's, and yet they are all yolk and that so thick that they must be diluted with water before they can be used in pancakes.

The slightly apocryphal tone of this reference suggests that the Great Auk was not abundant there in the late seventeenth century when Lahontan visited the region and, as we have seen, sightings of the bird in those regions a century later in Captain Cartwright's time were deemed to have been noteworthy. We can only guess about the former distribution of the Great Auk in North America, but it seems clear that Funk Island always held the greatest concentrations of the species in the breeding season. It is quite certain, however, that less than a century after Lahontan wrote there were few if any sites, other than Funk Island, where the Newfoundland Penguin could hope to rear its young. Cartwright, as a Labrador trapper, was aware of how vulnerable to over-exploitation the Great Auk had become by the mid-1780s, and concluded bleakly: 'This is now the only island they have left to breed upon, all others lying so near to the shores of Newfoundland they are continually robbed.'

Notes

1 This was no exaggeration: Spain's yearly return of American treasure peaked at £630 000 as early as 1545, declining to £280 000 by the end of the sixteenth century and progressively thereafter. In England, by contrast, the value of the sales of salted cod rose steadily, even if allowance is made for inflation: £220 000 worth in 1615, rising to £700 000 by 1640, and reaching as much as £800 000 in 1670.

2 Ascertaining the breeding times of different species is best carried out with reference to isothermic variation rather than to particular months of the year which may be quite different in temperature in different parts of the Northern Hemisphere. It is interesting to note, however, that the earliest reference to sea birds congregating at Funk Island preparatory to breeding is from the latter half of May 1534, when the ice had not fully retreated.

Certainly we should expect the Great Auk's breeding season to have varied considerably depending on either the latitude of its breeding grounds or, more especially, the presence or otherwise of a warm current in the adjacent seas. As with all birds, they would have timed their breeding to coincide with the availability of food for their young.

3 On his second visit to North America in 1535, Cartier kidnapped Donnaconna, the king of a Canadian Indian tribe, and took him back to France in the belief that he would, under pressure, reveal the route of the supposed North West Passage. This monarch, though treated kindly enough, died in captivity four years later, so it is not difficult to understand the military aspect of Cartier's third voyage in 1541.

4 Traditionally a French league was 4.448 km. However, the term has since been standardized with other countries so that later references can be understood as representing 1/20th of a degree, equivalent to 3.456 statute miles or 5.556 km.

5 The *th* is pronounced as *t*; the plural is 'Apponatz' — a name soon corrupted by mariners, both French and English, to 'Aponars'.

6 At first sight this seems to be a reference to Magpies — an unlikely choice as a standard by which to judge the size of other birds. Richard Hakluyt went one stage further and transcribed the word as 'Gaies' (Jays)! It seems likely that 'Pies' is simply a misprint for 'Oies', the French for geese which were, it is clear, widely used for the purposes of comparison.

7 This curious turn of phrase can be taken to mean that the birds used their wings as well as their feet to move rapidly through the crests of the waves when evading the ships.

8 Parkhurst gave 30–50 as the total of English ships going to Newfoundland annually, a figure which may be compared with 150 Breton, 100 Spanish (almost certainly from Galicia), and *c.*50 Portuguese; besides these fishing vessels, 20–30 Basque whalers also operated in the region.

9 It is possible that the reference to the 'Island of Penguin' in Thomas Butts' account of Robert Hore's voyage indicates this archipelago rather than Funk Island, for both lie north east from Cape Breton in Nova Scotia. Funk Island, however, is much the further of the two. A third possibility, however, is that Thomas Butts said, or meant to say, north-*west*wards, in which case the party might have arrived without undue delay at Bird Rocks in the Gulf of St Lawrence, the site which Cartier had visited on 25 June 1534 when he had found Great Auks and other sea birds in abundance.

A hearsay account, dated 1718, describes densely packed breedings birds on the Penguin Islands within living memory (see Montevechhi, W.A. and Kirk, D.A. (1996). Great Auk. In the *Birds of North America* No 260 (eds A. Poole and F. Gill). Philadelphia and Washington D.C.)

CHAPTER 6

Books of authority

Widespread ignorance about the extent of the Great Auk's former breeding range at the time of its extinction is not surprising given the superficial similarities of appearance which the penguin of the Newfoundland fishermen shared with its unrelated namesakes found on the rim of the Antarctic ice. However we may ask, how, with only circumstantial evidence in respect of the Great Auk's occurrence in the far North, such an assumption gained credence in scientific circles. Because ornithologists seldom used to set foot in the regions where the species actually bred, any misconception that appeared in print tended to go uncorrected and to be accepted as fact by writers on natural history subjects — be they in France, Great Britain, or North America. It is especially remarkable in the case of the Great Auk, however, that the misapprehension arose comparatively late and was far from being a primitive tradition which had simply gone unchallenged. It is unfortunate, and to some degree ironic, that a century before the Great Auk disappeared, the opinion of certain naturalists regarding its range was a much closer approximation to the truth than the considered opinion of even the best of the early nineteenth-century writers.

It is not difficult to envisage how it was that voluminous works, written expressly to incorporate every species known to science, soon acquired an almost hallowed reputation. No one in the nineteenth century felt the problems created by this more keenly than the British ornithologist John Wolley, a brilliant and energetic field naturalist. Writing in 1852 he observed:

> Assertions which find their way into books of authority are very long before they entirely lose credit. They are handed down from one writer to another; they are received as articles of early faith to which one is apt fondly to cling in after years: those who might make original observations not caring to run the risk of unsettling their former belief, whilst those who have no personal opportunities of enquiry prefer the established authority of their first favourite to that of anyone who has been rash enough to call it in question on any point.

The great encyclopaedists of the seventeenth and early eighteenth centuries relied on a range of sources, some of which were purely anecdotal and impossible to check. Moreover, many of the accounts received from sailors and missionaries to distant parts imparted an element of what we may term 'readability' to works which would otherwise have been compendia of dry-as-dust accounts. In this way, mariners' yarns were conflated with the first accurate reports on unfamiliar species received from local naturalists — often pastors or doctors — who had detailed knowledge of the areas where they lived and worked. As time went on the encyclopaedic authors increasingly yielded to the temptation to fill in the gaps in contemporary knowledge of a given species' natural history —a tendency which, we shall see, became particularly prevalent in the early nineteenth century. By contrast, the great naturalists who wrote prior to the middle of the eighteenth century, undiscriminating though they might have been in their selection of material for publication, were careful, generally speaking, not to elaborate their sources out of purely literary considerations.

Although Carl Linné (better known simply as Linnaeus) was the first to attempt a systematic arrangement of all known animal and plant species, he was not, of course, the first to be interested in the affinities of one species with another. In the late sixteenth and early seventeenth centuries, Charles de L'Écluse (known to his contemporaries as Carolus Clusius), a French writer[1] on the fabulous creatures discovered by European explorers, received information on large black and white sea birds from three different areas of the globe — near Cape Horn, the eastern seaboard of North America, and the Faeroe Islands — and accordingly gave descriptions of what he took to be three separate species. One of these was the 'Magellanic Goose or Penguin' first discovered by the Portuguese explorer Fernão de Magalhães (better known to posterity as Ferdinand Magellan) during his circumnavigation of the world in the early sixteenth century. He illustrated a second species, named simply 'Mergus Americanus' (i.e. 'The American Diver'), in the following terms in his *Exoticorum Libri Decem* (1605):

> A bird which is also a stranger and whose picture we give here: for that most distinguished man Jacques Plateau who sent me this clear picture in colour, wrote that it was brought from America and he was of opinion that it belonged to the genus *Mergus*, adding that he had received it from mariners.
>
> Truly, not one of those who has written on birds (who, it so happens, seem learned enough to me) has seen anything similar, and it is an unpleasant thing for me that in all my studies of exotica, I have not been able to work it out in view of the carefully drawn picture placed at the head of this section. As for the rest of its history I can offer nothing except what I have been able to put together from this picture alone, and certainly not anything added to that, which (I shall set down) exactly as these appear to me in the picture: what I now bring before the public I entrust to the gentle reader for his consideration.

Fig. 1 The earliest known picture of a Great Auk, from *Exoticorum Libri Decem* by C. Clusius (1605).

> The bird is a little less than a goose, or, rather, I understand that it is equal in size to a wild goose, furnished indeed with a long body but small and short wings not corresponding to the size of the body, on account of which as I adjudge, it is not very well adapted for flight: head, neck and back I conjectured to be covered with black feathers as depicted; also the plumes of the tail and wings, black; it is white, on the other hand, on the breast and the entire belly; the bill is aquiline and quite thick, not flat, in which no vestige of teeth appear evident; however, its forward part has several oblique striations, supposing the picture to be a faithful portrayal, and the front of the head a little above that part of the beak marked with a white spot — if indeed the illustrator drew that correctly; its legs appear to be short and black, the feet also black and flat after the fashion of web-footed birds, furnished with three toes and a short hind toe as in ducks.[2]

In 1604 de L'Écluse received a letter from a doctor by the name of Hoier, from Bergen in Norway, about the 'Goirfugel',[3] a visitor to the Faeroe Islands. De L'Écluse passed this information on to his readers with little elaboration, failing to notice the similarity that Hoier's description bore to his own of the 'Mergus Americanus'. Had he done so, he would have been able to establish at an early date that the range of the bird we now know as the Great Auk stretched right across the subarctic regions of the North Atlantic:

> The other is the Goirfugel, in colour not dissimilar to the body of a razorbill (Alka) even if it is greater in size; its bill is decidedly broad and down-curved; the head is rather oblong also black; the margins of the eyes are high-lighted with a gleaming white circle: it has black feet which are of no use for walking and in addition it has wings which are decidedly small. And in fact it has never been seen to walk or to fly either.
>
> The foregoing is very scarce (rarissime) and except in exceptional years it is not seen frequently: also let any explorer give his attention to the whereabouts of its breeding grounds.
>
> On both counts I enumerate this species in the genus *Mergus.*

As we might have expected it was not long before other naturalists began to see that there were correspondences between some of de L'Écluse's species. A Danish naturalist, Ole Worm (styled Wormius), writing in the mid-seventeenth century, considered that de L'Écluse's description of the 'Magellanic Goose' corresponded with specimens he had received from the Faeroe Islands. However, he did so evidently without having considered the earlier writer's descriptions of the 'Mergus Americanus' or the 'Goirfugel', for had he done so he must surely have recognized Hoier's description as fitting the species he had come to know well. In his volume entitled *Museum Wormianum* (1655), he merged his own description of the Great Auk with what he had read in de L'Écluse's work. Wormius' account is adorned with a lively illustration of a 'penguin' (i.e. a Great Auk) — the only one ever to have been drawn from life. The circumstances Wormius explained as follows:

> I shall set out the account of the Magellan's Goose or Penguin in the version of the most excellent Clusius, *Exoticorum* [&c.], and since I have three skins of this bird, one of which I kept alive at my home for some time, the things that I have additionally observed in that bird are not inaccurate.

Wormius then continued with a recognizable description of the *Spheniscus* Penguins from the Southern Hemisphere taken verbatim from de L'Écluse's account of the 'Magellanic Goose', to which were added some curious details of these birds 'making earths as conies do, in the ground [where] they lay their eggs and bring up their young'. This was presumably an inference drawn from the habit of certain species 'crèching' their young, found in *The World Encompassed*, an account of Sir Francis Drake's circumnavigation of the world in the late 1570s, published only in 1628. Wormius' account additionally includes one erroneous detail based apparently on the breeding plumage of the Great Northern Diver: 'The neck, which is short and thick, has a ring of white feathers like a collar' — a statement which goes much of the way to explaining why Wormius' illustration of the Great Auk appears as it does, although we can be certain that the bird which the naturalist had had in his possession showed no such feature. Wormius concluded with a brief summary of what he knew of the species from first hand observation:

> A bird was given to me from the Faeroes and for a few months I fed it in my home; it was a young bird because it had not reached the size whereby it would have been larger than a common goose. It was able to swallow a herring whole at one go and sometimes three in succession before it had filled its crop. The feathers of its back were so soft and even that it looked like black satin, its belly was of exceptional whiteness. It had a round white area above its eyes, about the size of a Thaler piece, and you would have sworn that it had been given spectacles (something that Clusius did not notice). Nor did the wings have the shape which he described, for they were a little wider, with a white edge. For this reason I have ensured that my bird was drawn from life so that the illustration might be the more accurate.

Unfortunately, Wormius seems to have given the bird in his illustration a curious hybrid form of the true wings of a Great Auk and the 'leathery appendages hanging down like two small arms'[4] described by de L'Écluse — just as, in fact, the former also inserted the white neck collar in an attempt to make the bird fit what he took to be the most authoritative description of it. Drawn from life, as his Great Auk certainly was, it is highly regrettable that it was not drawn simply on the basis of what he actually observed. This is not to say that the picture is wholly untrue to life — for instance, he captured the upright stance of the species better than any artist who illustrated it without having seen a living bird.

Wormius' very unscientific confusion of two unrelated *genera* can best be explained by reference to his belief that his tame bird was a young one — a belief which arose solely from his literal interpretation of the phrase 'the size of a Goose', found in the older descriptions — and his consequent wish to let the reader have an idea of what, in his opinion, the adult bird would look like. Whether this accords with the claim that 'the things I have additionally observed in that bird are not inaccurate', which he made in the final sentence of his account, seems more than a little doubtful.

After reading this lesson in the dangers of departing from scientific objectivity, it is refreshing to read an accurate account of the Great Auk on the Faeroes, also written in the second half of the seventeenth century. In *Ferroe & Feroa Reserata* (1672)[5] by Lucas-Jacobsen Debes, 'Provost of the churches there', we read:

> [H]ere cometh a rare waterfowl, called *Garfugel*, but it is seldom found on clifts under the promontories, it hath little wings and cannot fly; it stands upright and

Fig. 2 Wormius' Great Auk — the only picture ever published of a living bird. From *Museum Wormianum* (1655).

> goeth like a man, being all over of a shining black colour, except under the belly where it is white; it hath a pretty long raised beak though thin toward the sides, having on both sides of its head over the eyes a white round spot as big as a half-crown, showing like a pair of spectacles: it is not unlike the bird *pinquin*, that is found in Terra del Fugo [*sic*], painted and described in *Atlas Minor Mercatoris*. I have had that bird several times, it is easy to be made tame but cannot live long on land.

Wormius' illustration was reproduced on a smaller scale and with a shallower, less markedly decurved beak, in *The Ornithology of Francis Willughby* (1678) edited by John Ray (1628–1705). This, the first substantial bird book in the English language, was a somewhat expanded translation of the Latin text of Willughby's *Ornithologia* of 1676. Regrettably, and as a direct result of Wormius' confusion of the 'Garfugel' with the Southern Hemisphere Penguins, *both* kinds are treated by Willughby[6] under the heading 'Whole-footed birds that want the back toe', whereas in fact the *Spheniscus* Penguins have the normal complement. In this way the force of the opening statement — 'the bird called Penguin by our seamen which seems to be Hoier's Goifugel [*sic*]'[7] — loses much of the value that it would otherwise have had. However, Willughby does distinguish between 'the bird called Penguin by our seamen' and the 'Penguin of the Hollanders or

Magellanic Goose of Clusius' and he made the following significant remark which suggests that he thought the earlier naturalists might indeed have confused these two very different birds: 'Whether it hath or wants the back toe, neither Clusius nor Wormius in their descriptions makes any mention. In Wormius his figure there are no back toes drawn.'

We can see immediately that Ray's contemporary, disciple, and friend, Francis Willughby (1635–72), was a highly discerning ornithologist with an acute sense of what characteristics should be looked for in establishing the different genera of birds. It is most unfortunate, therefore, that he should ever have been held responsible for the confusion of the different flightless birds of the two hemispheres as he subsequently was by the writer and illustrator, George Edwards. After giving an accurate description of the 'Northern Penguin' in his *Natural History of Uncommon Birds* (1743–51), Edwards wrote:

> This bird I procured of a master of a Newfoundland fishing vessel who told me, it was taken with their fish baits on the fishing banks of Newfoundland near an hundred leagues from shore. This bird hath already been figured and described; but the figure has a ring round the neck in Willughby, which is not found in the natural bird and the descriptions are not clear; it is also confounded with the southern penguins, and Mr Willughby seems to think them and the northern the same birds; but I, who have seen several both from the south and the north, am so far from being of his opinion that I should rather make them of two distinct tribes of birds. The southern has four toes on a foot, though Mr Willughby says Clusius's figure is false in having four toes[8] whereas it is confirmed to me to be true. The southern also has different wings, and nothing on them to be called feathers.

As it is certain that Francis Willughby would have noted the difference immediately if he had had access to a specimen of a *Spheniscus* Penguin, it was uncharitable of Edwards to use the criterion the earlier ornithologist had established to question the accuracy of his account. Importantly, however, Edwards concluded with some accurate observations on the species' respective ranges:

> I have figured this bird principally to show that the above described bird is a distinct species, if not a distinct genus, from those called penguins about the Streights of Magellan and the Cape of Good Hope. The above described is a bird common to the northern parts of Europe and America, it being found in the island of Ferroe belonging to Norway.

Many of the later authors complimented Edwards on the accuracy of his plate and it is a pity that he did not produce a Latin edition of his *Natural History of Uncommon Birds* for, had he done so, more notice might have been taken of what he additionally wrote. It is remarkable that claims that the Great Auk was a bird of the *far* North can only be traced back to the period immediately after Edwards published his work.

Fig. 3 The 'Northern Penguin', from *The Natural History of Uncommon Birds* (1743–51), written and illustrated by George Edwards.

In seeking to explain how it was that a fundamental error came to be made at so late a date, it would probably be tedious, if not impossible, to try to trace it to its source. Instead we may observe how the belief, once accepted, was increasingly elaborated even in the working life of one naturalist. Taking Linnaeus as an example (not to cast aspersions on his reputation, but solely to trace the development of an idea), we see that in his *Fauna Suecica* (1746) he stated, perhaps with Hoier's testimony in mind: 'frequents Norwegian waters but rarely'. However, in the tenth edition of his *Systema Natura* (1758), he wrote: 'inhabits the European Arctic' — which is, most would agree, a significant shift of emphasis. In the 1788 edition of the same work, moreover, we find that Linnaeus wrote: 'inhabits the high seas of the European and American Arctic'.

In the meantime, the French naturalist, Brisson, whose *Ornithologia* was published in 1760 (see Chapter 1), had stated that the birds were found in northern waters — 'on les trouve dans les mers du Nord' — a phrase sufficiently vague to be open to a number of different interpretations and calculated, it almost seems, to excite the imagination. As all the eighteenth-century writers used to read each other's works,[9] there was probably extensive cross-fertilization of ideas. We may imagine that much was founded on the slenderest of evidence and built up to create a picture which was ultimately misleading.

Georges Louis Leclerc, Comte de Buffon, whose massive *Histoire Naturelle, Générale et Particulière* (1770–83) effectively eclipsed Brisson's work, qualified the received ideas about the Great Auk's supposed Arctic range by stating that 'the Faeroe Islands and the coasts of Norway seem to be their native land in the Old Continent just as Greenland, Labrador and Newfoundland are in the New' — a description which, apart from the reference to Norway, can be said to be broadly accurate but which gives no indication of the relative size of its breeding populations. We then find the introduction of what can only be described as a fanciful element into this account of the species:

> Penguins of the North and South, in spite of the characteristic differences which separate them, are oddities condemned to spend their lives in the company of equally strange beasts, imperfect and stunted shapes incapable of figuring with the more perfect forms in the centre of the tableau and exiled into the distance at the margins of the world.

Nothing could have been better devised to fix the idea in the reader's mind that the high Arctic was the true home of the Great Auk.

The Revd Otto Fabricius, a Danish clergyman who ministered in Greenland for many years, summarized the local status of the species in his *Fauna Groenlandica* (1780), a work which became well-known among the leading naturalists of the day: 'Inhabits the high seas, occasionally appearing on distant islands even in the winter season. Old birds are very rare.'

We may infer that by old birds ('veteres') Fabricius meant adults in breeding plumage, for at any distance young and old were indistinguishable once the large white loral spot of the mature bird was obscured by darker feathers after the autumn moult. This suggests immediately that the Great Auk was a visitor to Greenland during its post-breeding dispersal and that only late, lingering birds might be encountered in breeding plumage the following spring — an inference which is confirmed in Fabricius' unpublished manuscript *Zoologiske Samlinger* preserved in the Royal Manuscript Library at Copenhagen. It is strange, therefore, that Fabricius should have claimed to have examined the stomach contents of a recently hatched Great Auk from Greenland, and our credulity is severely tested when we read that he found in it only vegetable matter, including Rose-root (or 'Midsummer-men') *Rhodiola rosea*, rather than fish.[10]

Nothing illustrates better the principle that assertions once made 'are handed down from one writer to another', to quote John Wolley again, than the fact that a century and a half after Fabricius made his claim, it was included, apparently without so much as a raised eyebrow, in *The Handbook of British Birds*, Witherby *et al.* (1938–41). Other evidence which, more understandably, was taken to suggest that the Great Auk was a bird of the high North, includes Fabricius' reports that the local Eskimos used the skins of this species, as well as those of Guillemots, for making winter clothing and that the inflated stomach linings served as floats for their harpoon lines. The first of these two pieces of information, moreover, seems to have been corroborated by Thomas Pennant who wrote in his *Arctic Zoology* (1784): 'I think I have seen some habits of the Eskimaux made of the skins.' None of his contemporaries though seems to have remarked on the fact that Pennant never claimed to have encountered a single living Great Auk in the regions he described.

Far northern records of the Great Auk are very unusual and their reliability has often been doubted. The Danish ornithologist F. Benicken, writing in the *Isis* (1824), mentioned one bird said to have been taken in 1821 at Disko Island off the west coast of Greenland above the Arctic Circle. If authentic, this is the most northerly record of the species in recent historical times.

The most influential European bird book of the early nineteenth century was the *Manuel d'Ornithologie* by Coenraad Temminck (revised 2nd edn, 1820). Following the example set by Brisson, the detailed description of the bird is very good. As the early nineteenth century was strongly influenced by the Romantic movement in literature, perhaps we should expect to find the mood of the times reflected in reference works. Nothing, however, quite prepares us for Temminck's imaginative account of the range of the Great Auk, a bird which, he related, was found habitually *on the floating ice of the North Pole* (our italics) from which it distanced itself only accidentally.

The artist and ornithologist, John Gould, a frequent visitor to the Leyden Museum of which Temminck was the Director, seems to have lit his flambeau from the blaze of the Dutch ornithologist's prose. In his *Birds of Europe* (1837) he wrote:

> The seas of the polar regions, agitated with storms and covered with immense icebergs, form the congenial habitat of the Great Auk: here it may be said to pass the whole of its existence, braving the severest winters with the utmost impunity, so that it is only occasionally seen, and that at distant intervals, even so far south as the seas adjacent to the northernmost parts of the British Islands. It is found in abundance along the rugged coasts of Labrador; and from the circumstance of its having been seen at Spitzbergen,[11] we may reasonably conclude that its range is extended throughout the whole of the arctic circle, where it may often be seen tranquilly reposing on masses of floating ice, to the neighbourhood of which in the open ocean it seems to give a decided preference.

Gould was put upon inquiry by Japetus Steenstrup about his claim that the Great Auk was abundant in Labrador, but was unable to cite an authority unknown to the Danish professor. It became evident that the assertion rested solely on the 'admittedly vague and general tradition that this bird inhabited the coasts of Labrador and Greenland'.

Thomas Nuttall, a Yorkshireman who had become a naturalized American, displayed the same penchant for purple prose — in the florid style first adopted by Buffon — in his *Manual of the Ornithology of the United States and Canada* (1832–34): 'Degraded as it were from the feathered rank, and almost numbered with the amphibious monsters of the deep, the Auk seems condemned to dwell alone in those desolate and forsaken regions of the Earth.' Little did Nuttall or any of his contemporaries[12] realize, but the Great Auk had all but succumbed to a fate much worse.

Such was the state of knowledge of the Great Auk's range in the years immediately prior to its ultimate extinction — years when the clamour for specimens for both private and public collections was at its greatest. It was to take a further twenty-five years from the date of the publication of Gould's magnificent work for naturalists in the English-speaking world to realize that the supposed abundance of 'Northern Penguins' in Labrador and elsewhere in the frozen regions was a complete fallacy.

Notes

1 Clusius, who lived in the Netherlands, was a great botanist as well as a knowledgeable naturalist: his most enduring legacy has been the Dutch bulb and tulip industries which he pioneered.

2 The fact that the plate shows a bird with four toes, rather than three as it should have, was, of course, a lapse on the part of the artist.

3 It is possible that in giving the name as 'Goirfugel', Clusius had simply misread the word 'Garfugel', the Faeroese name, because of Hoier's handwriting.

4 '…loco alarum binas parvas coriaceas pinnas habet, in latera, tanquam parva brachia, propendentes…'

5 Taken from the English translation of 1676.

6 Francis Willughby had access to two specimens — one in the repository of the Royal Society, a second in 'Tradescant's Cabinet at Lambeth near London'.

7 '*Penguin* nautis nostratibus dicta, quae Goifugel Hoieri esse videtur.'

8 Francis Willughby said no such thing! All that he did say is given in full in this chapter.

9 We may compare the entry in John Latham's *Index Ornithologicus* (1790): 'Habitat in Europae et Americae borealis alto mari.' ('It inhabits the high seas of the European and American Arctic.')

10 Fabricius' somewhat contradictory conclusions were analysed by Prof. Japetus Steenstrup in his *Et Bidrag til Geirfuglens Naturhistorie etc.* (1857):

'For completion here is what Fabricius says in the *Fauna Groenlandica* (1780) p. 82 and what he adjudges to be the case in his remarkable manuscript *Zoologiske Samlinger*, part 1, p. 267: "One only sees this bird at the onset of winter from September to January, sometimes in great numbers but more often in small numbers; it is rare to see the old birds. They always keep themselves at a distance, rarely between the reefs and the coast and never on land. (Different observations could have been made without my knowing in the north of the colony.) In summer one does not see it as it stays in the localities where it nests — One cannot say that Greenland is its true home, as it does not nest on the coasts, and it does not approach except at considerable distance. I do not know where it nests as the Greenlanders have never seen its nest."'

'Immediately after Fabricius falls into self-contradiction as he states that once in the month of August he "saw a very young Alca which was far from its nest, still covered with down and only a few days old — This specimen could not, consequently, have come from a nest very far away."'

In this context Steenstrup cited Fabricius' account of this bird in *Fauna Groenlandica*: 'But I saw a chick, captured in the month of August, all covered in grey down, only a few days old; and in it I discovered *Rhodiola rosea* and other vegetable matter, which customarily grows on cliffs on the shore, but not, however, fish: this bird was not long out of its nest and of necessity it could not have come from far.'

By contrast, the citation from the unpublished journal continues: 'The Greenlanders of the district I inhabit (Frederikshaven) have traditionally been accustomed to visit all the isles by boat, even those furthest away, and they have never seen, in summer, either these birds or their nests. — In the western part of a big island, called Umenak, and further to the west in the reaches of the sea which are inaccessible, there only the *Alca* perhaps nests. There it probably lays its eggs in the midst of large rocks along the shore after the manner of petrels.'

'What is striking in Fabricius's observation,' continued Steenstrup, 'is that, for someone who knew the Greenlander and his love of nature and at the same time used his eyesight, it is difficult to believe that the natives had no cognizance of the nest of a bird of such great size. This last author therefore contradicts himself as he has assigned to a great distance the locality where one finds the nests, while at the same time alluding to the discovery of a young bird; there is no doubt but that there was a mistake about that on his part.'

A contrary opinion to that of Steenstrup has been put forward by Morten Meldgaard stressing Fabricius' reliability as an observer.

11 Whether the occurrence of the Great Auk at Spitzbergen was a matter of conjecture or rumour is unclear. Publication was first given to the notion by P.J. Selby in his *Illustrations of British Ornithology* (1825–33), but this writer later accepted that the information he had received had been given in error.

12 Jap Steenstrup was to visit Iceland in 1839–40 where he gained the impression from people he spoke to that there were still several Great Auks to be seen each spring.

CHAPTER 7

'Wild foulis biggand' – the Great Auk on St Kilda

For many years, until its evacuation in 1930, St Kilda was renowned as the most remote island community in the United Kingdom. Indeed, so difficult is it to reach that in spite of being the only place in Britain where the Fulmar Petrel could be found, until the relatively recent expansion in the range of that species, it was, then as now, seldom visited even by active ornithologists. One such, William Eagle Clarke, who stayed on the little archipelago in 1910, described the grandeur of the scenery of the main island, Hirta, in his *Studies in Bird Migration* (1912):

> The sides of the hills surrounding the crofts rise rapidly and when free from the masses of rough boulders and great screes, are clothed with short grass intermixed with which is a little thin heather. Crags and rocky faces present themselves on the higher parts of Connacher and Mullach Sgail. Beyond, to the north and east the summit rims of Oiseval, Connacher and their connecting ridges present to the Atlantic a line of magnificent cliffs — the finest in the British Isles, ranging from 500 to 1262 feet [150 to 385 m] in height.

One of the few ornithologists in the early nineteenth century with the zeal and determination necessary to reach the island was William Bullock who mounted an expedition there six years after his exploits in Orkney and Shetland (described in Chapter 1). He was amply rewarded for his pains with the discovery of the previously overlooked species of Petrel now known as Leach's Storm-Petrel. It is difficult to believe that the tradition, cited by Montagu in his *Ornithological Dictionary*, that the island was a breeding place of the Great Auk was not, in part, what had prompted him to go. With hindsight, however, we can see that it was not surprising that he had nothing to report on this species when he returned home. Since the middle of the eighteenth century, contemporary accounts of the breeding of the Great Auk on St Kilda had been based on no more than questionable hearsay, even if there appears to be a discernible core of truth behind them. Indeed, it may be doubted whether the species was ever observed there by anyone who may be rightly termed an ornithologist: the account, for example, given by the Revd Kenneth Macaulay in his *Voyage to and*

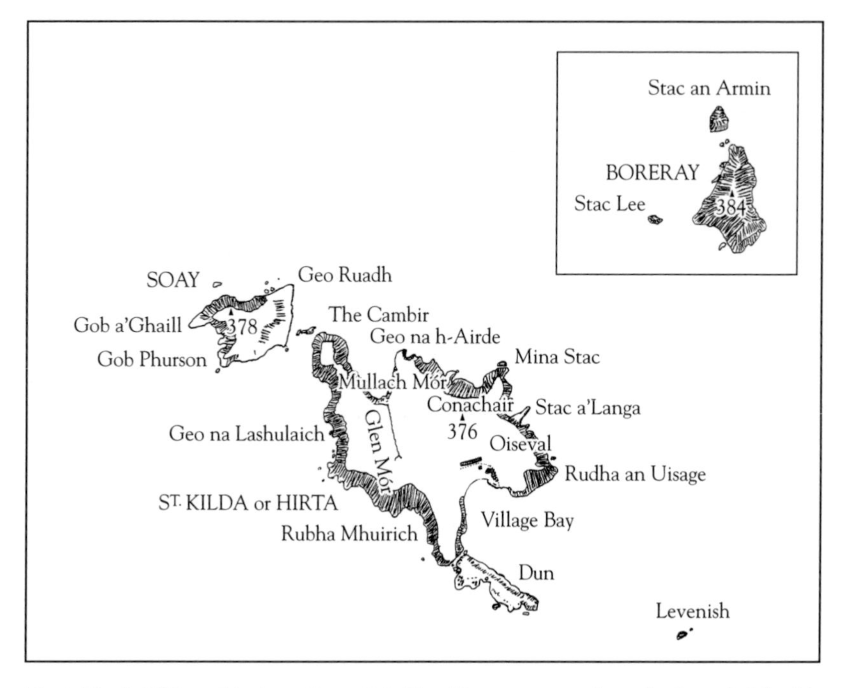

Fig. 1 The St Kilda archipelago, Outer Hebrides. The exact spot where the Great Auk bred is not known.

History of St Kilda (1764) contained a considerable degree of exaggeration regarding the bird's size and structure, together with some misplaced speculation on the part of the author:

> I had not an opportunity of knowing a very curious fowl, sometimes seen upon the coast, and an absolute stranger, I am apt to believe, in every other part of Scotland. The men of Hirta call it a *Garefowl*, corruptly, perhaps, instead of 'Rare Fowl', a name probably given it by some one of those foreigners[1] whom either choice or necessity drew into this secure region. This bird is above four feet in length from the bill to the extremities of its feet; its wings are, in proportion to its shape, very short, so that they can hardly poise or support the weight of its very large body. Its legs, neck, and bill, are extremely long; it lays the egg, which according to the account given me, exceeds that of a goose, no less than the latter exceeds the egg of a hen, close by the sea-mark, being incapable, on account of its bulk, to soar up to the cliffs. It makes its appearance in the month of July. The St Kildians [*sic*] do not receive an annual visit from this strange bird...It keeps at a distance from them they know not where, for a course of years. From what land or ocean it makes its uncertain voyage to their isle is perhaps a mystery of nature. A gentleman who had been in the West Indies [i.e. the New World] informed me that according to the description given of him, he must be the Penguin of that clime, a fowl that points out the proper soundings to seafaring people.

The exaggerated description of the size and proportions of the Garefowl in this account suggest that the stuff of legend had already begun to cling to this species in the island. The description of the size of the egg, on the other hand, is remarkably accurate if we accept that the 'goose' referred to is the Solan Goose, or Gannet, rather than the domestic kind.

The first reference by a British writer to the word 'Gare' as a name for the Great Auk is to be found in the section on birds in the *De Animalibus Scotiae* by Sir Robert Sibbald ('Physician in Ordinar and Geographer') published as a supplement to his *Scotia Illustrata* in 1684. Having written a fairly comprehensive account of the bird life of Scotland, this author added an appendix in which he listed various species on which he requested more detailed information.[2] At the head of the list stands: 'A bird called *Gare*, similar to a cormorant, with a very large egg'.[3]

In producing his treatise, Sibbald had elicited information from Sir George Mackenzie of Tarbat who, as Lord Register, had unrivalled knowledge of the Western Isles and what each of them produced. With this information Sibbald was able to compile a lively account of St Kilda:

> The island of Hirta, of all the isles about Scotland lieth furthest out into the sea, is very mountainous, and not accessible but by climbing. It is incredible what number of fowls frequent the rocks there, so far as one can see the sea is covered with them, and when they rise they darken the sky, they are so numerous. They are ordinarily catched in this way: a man lies upon his back with a long pole in his hand, and knocketh them down as they fly over him. There be many sorts of these fowls, some of them are strange shapes, among which there is one they call the *Garefowl* which is bigger than any goose and hath eggs as big almost as those of the Ostrich. Among the other commodities they export out of the island this is none of the meanest. They take the fat of these fowls that frequent the island, and stuff the stomach of this fowl with it, which they preserve by hanging it near the chimney, where it is dried with the smoke, and they sell it to their neighbours on the Continent [i.e. the mainland], as a remedy they use for aches and pains.
>
> Their greatest trade is in feathers they sell: and the exercise they effect most, is climbing of steep rocks; he is the prettiest man who ventures upon the most inaccessible, though all they gain is the eggs of the fowls, and the honour to die, as many of their ancestors, by breaking of their necks.

However many Great Auks once frequented St Kilda, the notion that there was a sufficient number to support an export trade is evidently an exaggeration. Fortunately, the manuscript notes of the information given by Sir George Mackenzie were deposited in the Advocates Library in Edinburgh and these indicate, as we might expect, that the truth of the matter was not especially remarkable: 'They take the fat of the fowls that frequent these places and stuff the stomach of the fowl with it...'.

That Mackenzie did not necessarily intend the Garefowl to be understood in this context is confirmed by another writer on St Kilda, Martin Martin, who visited the island in 1697, and who described the fat of the sea fowls, termed by the islanders 'giben',[4] as 'their great and beloved catholicon [i.e. 'cure-all'] with which they stuff the stomach of a Solan Goose, in fashion of a pudding'. It seems that Mackenzie's description of the Garefowl, with its egg as big almost as that of an ostrich, had caught Sibbald's imagination, something which seems confirmed by the omission of the names of the other birds that are mentioned in the manuscript notes — the 'Lavir' (or Guillemot), the 'Falkir' (or Razorbill), and the 'Gug' (as the islanders called the unfledged Gannet). It is not difficult to see why 'a bird called *Gare*' topped the list of species about which Sir Robert Sibbald wanted further particulars.

Well over a century before Sibbald's time, in 1549, the Revd (later Sir) Donald Munro, High Dean of the Isles, visited each of the Outer Hebrides in turn. Describing Hirta as 'a quite low-lying island ('Ane maine laiche ile') abundant in corn and grassing', he summarized its significant living creatures as 'fair sheep, falcon nests and wild foulis [i.e. fowls] biggand' — the last a word meaning well-built. Arguably this is the oldest reference to the Great Auk as a British bird, but as Munro's purpose was to give an accurate assessment of the wealth of each island, 'wild foulis biggand' is best taken as a term embracing all species sufficiently numerous to constitute a natural resource of considerable economic potential.

An anonymous author, probably also from the sixteenth century, cited in the Appendix to *Celtic Scotland* (1876) by W.F. Skene, described the life of the islanders in terms which would have been recognizable to generation after generation of St Kildans. We may discern, however, that this particular writer took a jaundiced view of their happy-go-lucky existence:

> *Irt* is a little isle of one mile long pertaining to [the head of the Macleod clan]. It is most fertile of sheep and fowls, whereof it pays one great matter yearly to the said Macleod and his factors. And albeit they use no ploughs but delve their corn land with spades, yet they pay yearly sixty bolls[5] victual. There is no horse nor mares in the isle, and but few nolt [cattle] to the number of sixty or thereby. There comes no men forth of this isle to [jousting] or wars, because they are but a poor barbarous people unexpert that dwells in it, using no kind of weapons; but their daily exercitation is most in delving and gathering their eggs, whereon they live for the most part of their food. They make no labour to obtain or slay any fishes, but gather some in the craigs albeit they might have abundance thereof otherways if they would only make labour therefore.

The same author remarks, however, that the people were devout and prompt in the payment of tithes for the maintenance of a clergyman — 'which they pay most thankfully and justly of any people'.

The best account of the inhabitants of St Kilda is to be found in Martin Martin's narrative of his *Late Voyage* to the island published in 1698. His long wished-for opportunity to visit St Kilda came about when the Revd John Campbell, Minister of Harris, was appointed on the recommendation of the Laird of Macleod to counteract the pretensions of a certain Roderick who, playing on the sweet simpleness of the people, claimed to be none other than St John the Baptist returned to earth to guide the St Kildans in religious matters.[6] Even from Martin's description of the crossing from Harris to St Kilda we receive a sympathetic portrait of the people whom this author describes as

> so plain and so little inclined to impose upon mankind, that perhaps no place in the world, at this day, knows instances like these of true primitive honour and simplicity; a people abhorring lying, tricks and artifices, as they do the most poisonous plants or devouring animals.

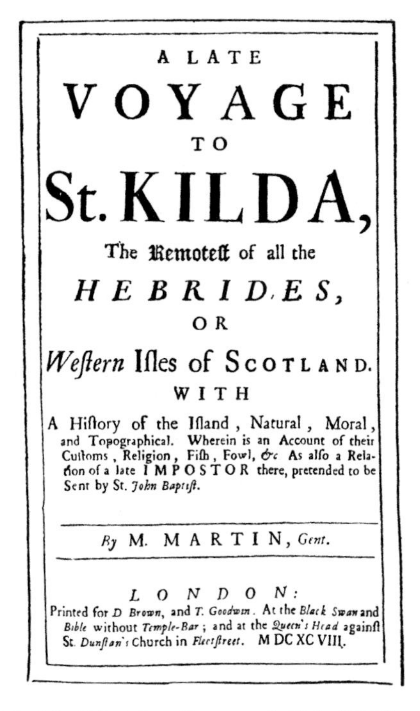

A LATE
VOYAGE
TO
St. KILDA,
The Remoteſt of all the
HEBRIDES,
OR
Weſtern Iſles of SCOTLAND.
WITH
A Hiſtory of the Iſland, Natural, Moral, and Topographical. Wherein is an Account of their Cuſtoms, Religion, Fiſh, Fowl, &c As alſo a Relation of a late IMPOSTOR there, pretended to be Sent by St. *John Baptiſt*.

By M. MARTIN, *Gent.*

LONDON:
Printed for *D. Brown*, and *T. Goodwin*. At the *Black Swan* and *Bible* without *Temple-Bar*; and at the *Queen's Head* againſt St. *Dunſtan's* Church in *Fleetſtreet*. M DC XC VIII.

Fig. 2 Title page of Martin Martin's *A Late Voyage to St Kilda* (1698).

Departing on 29 May 1697 — which, allowing for calendar changes which took effect in 1753 is about the 9th of June — they soon ran into difficulties with a strong wind blowing up from the south-east and had an anxious time of it. The next day the crew was:

> ...extremely fatigued and discouraged without sight of land for sixteen hours; at length one of them discovered several tribes of the fowls of St Kilda flying, holding their course southerly of us, which (to some of our crew) was a demonstration we had lost our course by the violence of the flood and wind both concurring to carry us northerly, though we steered by our compass right west. The inhabitants of St Kilda take their measures from the flight of those fowls when the heavens are not clear, as from a sure compass, experience showing that every tribe of fowls bend their course to their respective quarters, though out of sight of the isle; this appeared clearly in our gradual advances; and their motion being compared did exactly quadrate with our compass. The inhabitants rely so much upon this observation that they prefer it to the surest compass; but we begged leave to differ from them, though at the same time we could not deny their rule to be as certain as our compass. While we were in this state we discovered the isle Borera, near three leagues north of St Kilda, which was then about four leagues to the south of us; this was a joyful sight and gave new vigour to our men, who being refreshed with victuals, lowering mast and sail, rowed to a miracle: while they were tugging at the oars, we plied them with plenty of *Aqua vitae* to support them, whose borrowed spirits so far wasted their own, that upon our arrival at Borera, there was scarce one of them able to manage cable or anchor: we put in under the hollow of an extraordinary high rock to the north of this isle, which was all covered with a prodigious number of Solan Geese hatching in their nests...
>
> We proposed being at St Kilda the next day but our expectation was frustrated by a violent storm which almost drove us to the ocean; where we incurred no small risk, being no ways fitted for it; our men laid aside all hopes of life, possessed with the belief that all this misfortune proceeded from the impostor Roderick who they believed had employed the Devil to raise this extraordinary storm against Mr Campbell who was to counteract him. All our arguments, whether from Natural Reason or the Providence of God, were not of force enough to persuade them to the contrary, until it pleased God to command a calm the day following which was the 1st of June, and then we rowed to St Kilda. As we came close upon the rocks, some of the inhabitants, who were then employed in setting their gins, welcomed us with a 'God save you!', their natural salutation, admiring to see us get thither contrary to wind and tide; they were walking unconcernedly on the side of this prodigious high rock, at the same time keeping pace with our boat, to my great admiration, in so much that I was quickly obliged to turn away my eyes, lest I should have had the unpleasant spectacle of some of them tumbling down into the sea; but they themselves had no such fears, for they outran our boat to the town, from thence they brought the steward and all the inhabitants of both sexes to receive us.

Fortunately for posterity, no aspect of St Kildan life was without interest to Martin Martin. He even included in his account a list of the land birds[7] on St Kilda, noting that the Cuckoo 'is said very rarely to be seen here, and that upon extraordinary occasions, such as the death of the proprietor Macleod, the steward's death or the arrival of some notable stranger'. In this we see similarities between the St Kildans and the inhabitants of both Iceland and the Faeroes who likewise saw the appearance of birds of only irregular occurrence as omens. Particularly notable in Martin's accounts of the sea birds are his descriptions of the Gannet, including detailed measurements and observations of nesting birds, and that of the Fulmar, both of which bettered anything produced in the next hundred years. This accuracy of observation has a direct bearing on how we should assess a description of what for Martin, as for Sir Robert Sibbald in the preceding decade, was the most distinctive of St Kilda's breeding birds:

> The Sea-Fowls are, first, Gairfowl, being the stateliest, as well as the Largest of all the Fowls here, and above the Size of a *Solan* Goose, of a Black Colour, Red about the Eyes, a large White Spot under each Eye, a long broad bill; stands stately, its[8] whole body erected, its Wings short, it flyeth not at all, lays its egg upon the bare Rock, which, if taken away, it lays no more for that year; it is *Palmipes*, or Whole-Footed, and has the Hatching spot upon its Breast, i.e. a bare Spot from which the Feathers have fallen off with the heat in Hatching; its egg is twice as big as that of a *Solan* Goose, and is variously spotted, Black, Green,[9] and Dark; it comes without regard to any Wind, appears the first[10] of *May*, and goes away about the middle of *June*.

Opinion has always been divided on whether this account was based on personal observation, which some of the detail and the repetition of the Gairfowl's 'stateliness' suggest that it was, or from what he was told by the islanders. Ultimately it matters not, for, as Martin stated in his Preface, he had been careful 'to relate nothing in the following account but what he asserts for truth, either upon his own particular knowledge, or from the constant and harmonious testimony given him by the inhabitants'. We have reason to be thankful that Martin Martin was such a painstaking author, for his laconic description of the Gairfowl remains the best published account of the Great Auk on its breeding grounds from anywhere within its extensive range across the north Atlantic.

During the following century another visitor to St Kilda, probably a relative of the MacLeod family who owned the island, wrote the following account of the Great Auk, which while brief, accords with the testimony of Sir George Mackenzie of Tarbat and Martin Martin:

> The Gernhell [*sic*] is the most remarkable fowl about St Kilda, for his enormous size and rarity, his wings are so very small in proportion to his bulk that he does not fly: they are taken by surprising them where they sleep, or by intercepting their way to the sea and knocking them on the head with a staff: they lay their eggs a little above the sea mark on rocks of easy access: they carry off their young soon to feed them at sea.

We can be sure that there was nothing haphazard about the way in which the various sea birds[11] and their eggs were exploited. According to Japetus Steenstrup (1857) the cliffs were:

> ...divided up according to a scheme by which the collecting of the birds could be carried out more easily. For example, several areas yielded eggs in the first weeks [of the breeding season]; during this period the undisturbed birds, in the other areas, got on with incubation — the young could be taken later — and the birds went on to nest a second time in the first locality to be exploited. There was time enough for this during the summer and, according to species, birds were left to incubate for a longer or shorter period of time, in order for the young to grow to a big enough size and so to be collected in the most advantageous way. The controlled exploitation of the rocks dictated accordingly, in the light of the number of young and eggs taken, how many of the old birds it was necessary to leave.

Steenstrup conjectured that the population of Great Auks on St Kilda was localized and that it was inadvertently extirpated by over-exploitation during an epidemic when the women of the island bore the responsibility for collecting eggs, concentrating their efforts on ground-nesting species. However, this theory does not take into account the longevity of larger species, a factor which would have made it necessary for such a process to be repeated year after year before a local population died out. We may question, additionally, the assumption that the population was completely self-contained: it is highly likely that the Great Auk became increasingly irregular on St Kilda just as its sudden appearances in considerable numbers off Iceland and the Faeroes became more infrequent. It is well known that a species in decline will tend to concentrate in its heartlands and recolonize the periphery of its range only when its numbers have built up to something like their former strength. It is likely, too, that the later records of the Great Auk on St Kilda refer to birds appearing during the post-breeding dispersal. This would account for the discrepancy between the dates given for its usual appearances by Martin and Macaulay and for the bird taken into care by Dr Fleming (see Chapter 1) being captured in August.

Martin Martin's reflections on his trip to the island suggest that what he had encountered there exceeded all his expectations:

> The inhabitants of St Kilda are much happier than the generality of mankind, being almost the only people in the world who feel the sweetness of true liberty...They are altogether ignorant of the vices of foreigners and governed by the dictates of Reason and Christianity as it was first delivered to them by those heroic souls whose zeal moved them to undergo danger and trouble to plant religion here in one of the remotest corners of the world.
>
> There is only one thing wanting to make them the happiest people in this habitable globe, viz. that they themselves do not know how happy they are, and how much they are above the avarice and slavery of the rest of mankind. Their way of living makes them contemn gold and silver as below the dignity of human nature; they live by the munificence of Heaven, and have no designs upon one another, but such as are purely suggested by justice and benevolence.

> They are very charitable to their poor, of whom there are not at present above three, and these carefully provided for by this little commonwealth; each particular family contributing according to their ability for their necessities; their condition is inquired into weekly, or monthly, as their occasions serve; but more especially at the time of their festivals.[12] They slay some sheep on purpose to distribute among the poor, with bread proportionable; they are very charitable likewise to strangers in distress; this they had opportunity to express to a company of Frenchmen and Spaniards who lost their ship at Rocol [i.e. Rockall] in the year 1686 and came in, in a pinnace to St Kilda where they were plentifully supplied with barley, bread, butter, cheese, Solan Geese, eggs, etc.

So things had continued since time immemorial and so they would continue for two and a half centuries yet. In due course, however, the siren voices of modern civilization were to echo alluringly about the craigs and storm-drenched gullies of this, the most remote of the Hebridean islands. Appropriately it fell to the writer and naturalist Seton Gordon to compose a lament, laconic enough in truth, for the passing into history of a unique community — an event which he felt could have been forestalled if the National Trust had owned the little archipelago then as they do today. In the 1950s Seton Gordon made his first visit to St Kilda in more than thirty years. Recalling a happier occasion he wrote:

> Then Hirta was still inhabited and as the *Dunara Castle* dropped anchor I could see, through my glass, the entire population of Hirta, the women wearing red shawls, entering the small church for afternoon service that Sunday. A few years

Fig. 3 A St Kilda fowler at work.

later the native population at their request were transferred to the mainland of Scotland. The younger generation found the life too lonely, and the older people felt that they could not carry on without them.

Notes

1 Macaulay had in mind the person(s) whom he supposed to have been responsible for naming the island after St Kilda, an Anglo-Saxon saint.

2 'Quarum proinde Descriptiones accuratas desidero.'

3 'Avis *Gare* dicta, corvo marino similis, ovo maximo.'

4 'Giben', coincidentally or not, happens to be the Arabic for cheese.

5 1 boll is the equivalent of 6 bushels or 48 gallons.

6 Eventually this Roderick was discredited when overheard and discovered in the act of attempting to seduce the wife of one of the islanders.

7 'Hawks [i.e. Peregrine Falcons] extraordinary good, Eagles, Plovers, Crows, Wrens, Stone-chacker [i.e. Wheatear], Craker [i.e.Corncrake], and Cuckoo.'

8 In a brief list of 'Errata' at the end of the volume, Martin's printers asked the reader to substitute 'it' and 'its' in this paragraph for 'he' and 'his' which are to be found in the actual text.

9 Green is said to be very unusual on the egg of this species, but it is not unknown (see Appendices).

10 i.e. 'Old May Day'. Following the calendar changes in the middle of the eighteenth century, eleven days need to be added to find the equivalent date today.

11 The other sea birds, in the order in which Martin gave them are: Solan Goose, Fulmar, Black Guillemot, Guillemot, Razorbill, Puffin, Storm Petrel, three sorts of 'Gull' (a term which seems to include the Arctic Tern), and, lastly, Oystercatcher. These were all given under their local names.

12 These were: Christmas, Good Friday, Easter, St Columba's Day, and All Saints.

CHAPTER 8

The New-found-land

In the years after Sir Humphrey Gilbert's expedition to Newfoundland in 1583, increasing numbers of fishermen from the south west of England began making annual expeditions to the Great Bank. Nearly a century earlier, John Cabot had persuaded Henry VII, King of England, that a short cut to the Pacific Spice Islands lay to the north on the grounds, rightly enough, that distances across the globe are shorter further from the equator. On returning from his voyage in 1497, Cabot had reported that the waters there were so rich in cod that baskets could be filled with them merely by being lowered into the water. Indeed, it was said, the cod were so many that the progress of his ship was impeded. Newfoundland, and the islands round about, he named 'Baccalaos', meaning, in effect, 'Codland' — a name which owes its origins either to the Basque language or to that of the Beothuk Indians — and five centuries of no-holds-barred exploitation thereupon commenced.

The manner in which the fishermen worked and the primitive conditions in which they lived changed little over the years. The account given in *The American Gazetteer* (1762), to which reference has previously been made, shows clearly what 'following the Newfoundland fishery' would have meant for those who yearned for the prospect of good money:

> When a ship has taken her station [in the spring] she is immediately unrigged; and at the same time a proper place chosen for securing the fish, as it is prepared; huts are likewise run up for the men to work ashore, so as to form a kind of village; and at the water's edge a large stage or scaffold is erected. There the number of shallops destined for the fishery is got ready, and when the season is over, left there until the next year; when he who first encounters the bay has the privilege of applying them to his own use…The fishers go out early in their boats, that they may be at their station by break of day, and do not return till the evening, unless they happen to have loaded their boat before. This fishery is only carried on with the hook; and every boat is provided with a sufficient quantity of fishing tackle in case of any accident in breaking their lines, or losing their hooks…As the boats go constantly every day, the work of the several classes may be imagined pretty hard and fatiguing. On the return of the boat [those on shore]

> immediately begin opening and salting the fish which takes up the greater part of the night; and the succeeding parts of the cutting necessarily employs them the following day when the return of the shallops causes them to renew their tasks; so that they have very few hours left for sleep and refreshment.

Examination of a large-scale map of Newfoundland reveals, by the curious place-names of the different headlands and bays (such as Port au Basques, Point Enragée, and Avalon Peninsula), the nations from which the first generations of fishermen came and the parts of the coast that each frequented most. As Cabot had put into the harbour of what would become the island's capital on June 24th, which happens to be St John the Baptist's day, he named the haven accordingly. The adjacent parts of the island became the centre of English colonization, and if this region was closer to the Great Bank than the settlements of other nations, it was simply a matter of good fortune. Placentia Bay, in the south west, became the principal French-speaking part of the island. Nearby, Fortune Bay was visited by the Revd Edward Wix in 1835 who left the following account of inshore[1] fishing during a Newfoundland winter in his *Journal* published the following year:

> They continue catching fish till Christmas, when the fish generally failing for a season, they avail themselves of this respite, to do their winter's work in making boats etc. They begin fishing again at the latest by Lady Day [March 25th]. It is exceedingly deep water in which they fish, by which the labour is much increased. The fishing-lines freeze as they draw them out of the water; after the first fish is caught, they throw them into the water coiled, that they may thaw in the sea. I have myself seen the fish, as soon as they have been taken out of the water, turn up from the cold and die immediately, stiff-frozen, and could not but pity the poor men who were subject to such exposure in rough weather.

Only gradually had the idea of remaining on the island suggested itself as a practical alternative to the weeks spent crossing the Atlantic to and from the fishing grounds each spring and autumn. For those who rejoined their families at home during the winter, the crossing the following year was something of a race, the prize for the first arrivals being the right to control the affairs of the harbour or creek where they took up station as well as to make use of the equipment already there. For many years these men were known as 'fishing-admirals' and we catch a glimpse of their determination to be the first to make landfall in the diary of James Yonge, a ship's surgeon from Plymouth who visited the island in April 1663: 'Those mad Newfoundland men,' he wrote, 'are so greedy of a good place they ventured in strangely'. We may wonder what kind of person it was who preferred to spend the winter season shivering in the frozen harbours rather than joining the exhilaration of this annual race.

In 1618, Captain Richard Whitbourne of Exmouth visited Newfoundland on behalf of certain patentees — as those who had obtained mineral exploitation rights from the crown were termed — and encountered 'such idle persons' as he 'had never known to be sober' and who 'never applied themselves to any commendable thing, no not so much as to make themselves a house to lodge in, but lay in such cold and simple rooms all the winter as the fishermen had formerly put there for their necessary occasions the year before'.

Whitbourne had conceived a scheme for a planned settlement — a 'plantation' — of Newfoundland with the intention, in part at least, that 'the poor misbelieving inhabitants of that country' (by which he meant the Indians):

> ...may be reduced from barbarism to the knowledge of God and the light of his truth, and to a civil and regular kind of life and government...and...it is not a thing impossible but that by means of these slender beginnings which may be made in Newfoundland, all the regions adjoining thereunto may in time be converted to the true worship of God.

Whitbourne outlined how his ideal might be realized in practice in his *Discourse and Discovery of New-found-land* (1620), the full title of which continues: '*with many reasons to prove how worthy and beneficial a plantation there may be made, after a far better manner than now it is together with a laying open of certain enormities and abuses committed by some that trade to that country and the means laid down for reformation thereof*'. Two years later he followed this with a sequel in the same vein: '*Discourse containing a loving invitation both honourable and profitable to all such as shall be adventurers either in person or purse, for the advancement of his Majesty's most hopeful plantation in Newfoundland lately undertaken*.'

The flaw in Whitbourne's scheme, as for all such plans for the colonization of Newfoundland, was that the terrain was unsuitable for agricultural development. Consequently, and rather in the manner of Anthony Parkhurst forty-five years earlier (see Chapter 5), Whitbourne stressed the abundance of wildfowl which, in his view, made the prospects for the self-reliant colonist distinctly promising even if the land itself could not be said to be flowing with milk and honey. Indeed, the workings of a beneficent providence were not difficult to read even in the annual return of countless sea birds, including of course the Penguins, to their ancestral breeding places. In the second of his tracts he told his readers: 'These penguins multiply so infinitely upon a certain flat island that men drive them from thence upon a board into their boats by hundreds at a time as if God had made the innocency of so poor a creature to become such an admirable instrument for the sustenation of Man.'

The abundance of sea birds and their eggs was noted by Whitbourne's contemporary, John Mason, in *A Brief Discourse of the New-found-land* (1620). His theme, if expressed with rather less sententiousness, was similar: 'The sea fowls are gulls white and gray, penguins, sea pigeons, ice birds, bottle noses, with other sorts strange in shape, yet all bountiful to us with their eggs as good as our turkey or hens, wherewith the islands are well replenished.'

So impressed was King James I with Whitbourne's scheme for a renewed attempt at colonization that he ordered that a copy of his tract be publicly displayed in every parish in England. Unsurprisingly, perhaps, things did not turn out as happily as the admirable Captain Whitbourne hoped. Those known as venturers, who did respond to the 'Loving Invitation', resented the presence of 'squatters' on the coast, who were in the habit, among other things, of denuding the shoreline of trees in their quest for firewood. In turn, the venturers were seen as interlopers. To put a no-man's-land between the two groups, the Government decreed that no new settler should build within six miles of the coast — though there were no means by which this edict could be enforced. Finding it impossible to wrest a living from the soil alone, many of the venturers were faced with the choice of returning to England or moving with their families to join the communities of those very 'squatters' they despised. It is likely that a number of them felt that Whitbourne had misrepresented what conditions were really like in the country.

Newfoundland's 900-mile coastline — over 2000 miles if you explore every creek, bay, and inlet — and the lack of communicating roads at that time between the different fishing stations would, under even the best of circumstances, have rendered effective government very difficult. In contrast to the French crown authorities, who from the earliest days had insisted on a centralized 'seigneurial' system of government in North America, the English Government seems to have been prepared to leave the day-to-day administration of the communities, which flourished chiefly in the summer, to the so-called 'fishing-admirals'.

We can well imagine that, in spite of the high hopes for the intended plantation, wrongs were committed which could all too easily go without redress. As early as 1630, the Government of Charles I observed, in its 'Commission for the well-governing of his [Majesty's] subjects inhabiting Newfoundland' that those who resorted thither 'injure one another and use all manner of excess to the great hindrance of the voyage and common damage of this realm', noting that these subjects, some of them probably people who had good reason to 'lie low' for a while, 'have imagined that for wrongs or injuries done there, either on the shore or in the sea adjoining, they cannot be here impeached'. Steps were taken in consequence to ensure that the Assizes and Quarter Sessions in Exeter and other centres of population in the West Country could try anyone for felonies and misdemeanours committed on the other side of the Atlantic. Increasingly, the type of person who stayed on in Newfoundland did so because he had reason to fear returning home.

In spite of the abundance of fish offshore during the summer there was an ever-present fear of want in the winter months. Some, both English and French, undertook trapping expeditions, either to supplement their incomes or as an alternative means of making a livelihood. This brought them into conflict with

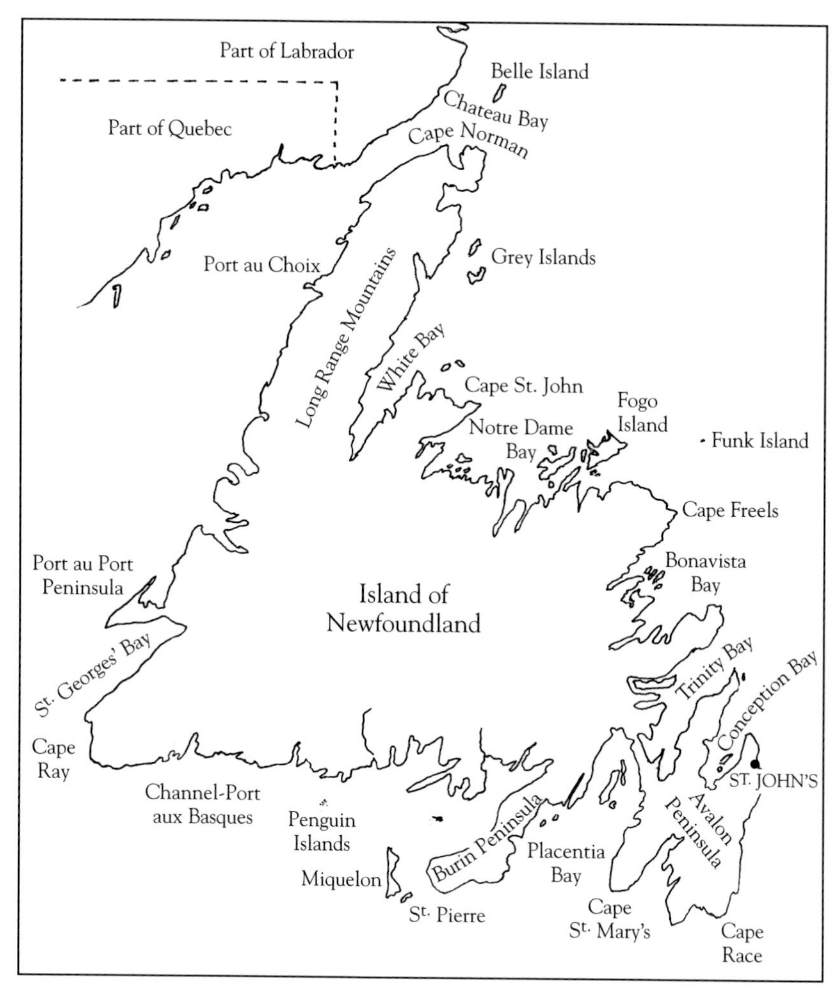

Fig. 1 Map of Newfoundland's 900-mile (1450 km) coastline showing the position of Funk Island and Penguin Islands.

the Beothuks in the interior who were intimidated and even murdered for their valuable furs which they were reluctant to part with in trade. For those who remained on the coast, provision for a twelve- or sixteen-week winter was imperative when most of the salted cod went to enrich those merchants of south-western England who financed the fishing expeditions. The following manuscript fragment relating to a colony known as 'Cupids' on the north-east coast of Newfoundland dates from 1613; it is representative of any year in the seventeenth century — an era characterized by a decided scarcity of records relating to the Great Auk: 'We are gone all abroad a-coasting all the [islands] for Eggs and birds against [the winter] which in one Island [to] the Northwards they may fill [the boats] with penngwynes.'

These depredations would have occurred all round the coast and the breeding population of Great Auks on such archipelagos as the Penguin Islands off Cape de la Hune in south west Newfoundland (mentioned in Chapter 5) in all probability died out[2] before the end of the seventeenth century as a result of over-exploitation by the adjacent French-speaking communities.

Whereas in the early days the fishing off Newfoundland was a free-for-all, increasingly France and England developed differing strategic interests. Once the volume of trade between Newfoundland and New England became significant, it was no longer possible to view the fishing communities, and the annual expeditions to them from the south-west of England, in isolation. The Government in London needed to take steps to ensure that the economic benefits of peopling the New World did not drain the mother country. As early as 1634, England, in pursuance of its claim to sovereignty over Newfoundland dating back to the time of Sir Humphrey Gilbert, enforced a payment of tribute on France for the right of her seafarers to fish there. In time it became evident to both nations that the waters off Newfoundland produced skilful mariners as well as fishermen and that each country's status as a sea power was to some extent dependent on this continuing. The purely strategic value of Newfoundland to the English nation was summed up by the writer John Ogilby in his *America* (1671):

> By well planting and fortifying Newfoundland the trading to Virginia, New England, and those parts, would be much encouraged, New England having had of late great traffic with Newfoundland, where they vend the growth of their plantation. Besides, Newfoundland is a key to the Gulf of Canada which if the English had again in their possession as they had in the years 1628 and 1632 they might give a law to all foreign kings and people interested in any of the parts of America and a protection to all the English plantations upon that continent, whereby great security, comfort and profit would arise to the nation and people of it, whereas, on the contrary, if the French, Spanish or Dutch should possess themselves of the said plantation, they would not only deprive this nation of all the advantages aforesaid, but would also derive to themselves so great a power to prejudice all the plantations of the coasts of America belonging to this nation, that the inconveniences thereof are scarce to be imagined or expressed.

There was in fact little that the crown could do. In order to prevent the fishing from becoming the exclusive fiefdom of a small self-interested group of merchants and fishermen, the Government had insisted that each fishing vessel departing in the spring had on board at least one crew member — a 'greenhorn' — not yet trained, hoping by this means that the seas between St John's and the Great Bank would become 'nurseries for our seamen'; strict account, moreover, had to be kept of the number of boats in the havens, of the produce taken there, and of the number of men employed, it being the responsibility of the fishing-admiral to deposit the requisite information with the Privy Council on his return to England. In 1697 these requirements were given statutory force, and the power and authority of the fishing-admirals ('according to the ancient custom there used') was confirmed. Section 16 of this Statute of William III, as it became known, imposed Lord's Day observance on the people there — though with what hope of either acquiesence or enforcement it is difficult to say. Most likely, it was a symbolic gesture towards maintaining the ideal of a Christian civilization and not much more.

The year 1697 was also, coincidentally, the first in which West Country merchants and shipowners lobbied for the appointment of a Governor, fearful that the better organized French settlers in Newfoundland might secure control of the entire island — as they had tried to do the previous year, when they had destroyed every English settlement there. For its own good reasons[3] the Government neglected to appoint one until 1729, and then only a naval Governor of the rank of captain or commodore, whose role did not extend beyond escorting the fishing fleet out in the spring, remaining in St John's for the season, and returning home in the autumn. In the meantime, twice during the War of Spanish Succession in the first decade of the eighteenth century, the French Governor in Placentia had repeated his outrages with apparent impunity.

Some kind of justice, rough enough at best to be sure, had always resulted from the supervision of the fishing-admirals. But we may well wonder what happened after the fishing fleets departed at the end of the season. The anarchy which then ensued was outlined in 1715 by a naval officer, Captain Kempthorne, in a letter to the Admiralty by whom he was engaged both to escort the fishing vessels and to visit the Newfoundland communities from time to time to ensure their well-being:

> The winter season is a rest of respite from all observance of law. At that time, theft, murder, rapes, or disorder of any kind whatsoever may be committed — and most of them are committed — without controul and time enough given for the party to make off: for should anyone concern himself to secure the party, his design would be withstood, as an usurped authority; and most would take part with the offender, to suppress the usurpation without regard to what became of the criminal, or what may be the consequence of the crime; ...there seems an absolute necessity...that people may always have somebody to apply to for justice; that somebody may always be at hand to suppress disorder and riot and to have a lawful power to command the assistance of his Majesty's subjects in the execution of a duty exercised for the public good...I do not know anything that tends more to confusion and proves more prejudicial to the fishery, than that irregularity.

In support of his argument for the appointment of a Governor, Captain Kempthorne was at pains to point out that matters were far from perfect even at other times of the year. The chief cause of complaint was the arbitrary way in which debts were collected, the debtor often being left without the means of securing a living the following season, so that he had to hire himself out to another. The point was also taken up by another writer of those times who commented:

> The [fishing] admirals prove generally the greatest knaves, and do most prejudice, being both judge and party, in hearing suits for debt; and when they have saved themselves, then they will do justice to others: so it would be requisite to have a civil government, and persons appointed to administer justice, in the most populous and frequented places, that they may be governed as Britons, and not like a *banditti* or forsaken people, without law or gospel, having no means of religion, there being but one clergyman in all the country.

These sentiments were, if anything, expressed more forcibly by Captain Kempthorne who continued:

> On the foot it is now on, he that happens to be the strongest knows everything to be his own, and the weakest knows nothing, or had as good as know nothing, except in that little interval of time when his Majesty's ship, or ships, happen to be there; and very often the aggressor absconds, runs into the woods, and flies from justice until the ships are gone; and then down he comes and reigns lord again.

He concluded his trenchant criticisms in the frankest terms, expressing the view 'that no man living in the country of Newfoundland is fit to govern. For the set of people that live here are those that cannot live in Great Britain, or anywhere else, but in a place without government.'

Notes

1 Wix describes fishing on the so-called 'French shore', an extensive stretch of coastline to which French fishermen had exclusive fishing rights since the eighteenth century (see Chapter 9).

2 Two eggs discovered at the Lyceum, Versailles, and acquired by the Museum of Natural History, Paris, in 1873 were found to be inscribed 'St Pierre, Miquelon' — the French possession off the south west tip of Newfoundland. As the eggs presumably do not date from before the late eighteenth century, it seems that small numbers of Great Auks still bred on accessible islands at some distance from Funk Island until a later date than has generally been supposed.

3 Notwithstanding the War of Spanish Succession, British foreign policy towards France, and indeed other European powers, in the early eighteenth century was very much based on the principle of 'live and let live'; as a result it was considered expedient to interpret this kind of violence as part of a

conflict over settlement rights rather than a bid to oust the English colonists altogether. A much more aggressive foreign policy was adopted by the Earl of Chatham, better known as Pitt the Elder, who entered the Government in 1746.

CHAPTER 9

Uncouth regions

The naval Governors who were appointed to administer Newfoundland from 1729 onwards, typically for terms of three successive fishing seasons, must have regarded such a posting as no more than a step to advancement and chafed at the lack of any realistic prospect of action. In view of the intractability of the island's social problems and the comparative brevity of each Governor's term, it is not surprising that there was little or no attempt to effect a reformation in the behaviour of Newfoundlanders until the last quarter of the eighteenth century. At that time the problems deriving from the absence of any elementary education were at last addressed and, in addition, a Court of Common Pleas was established in which to try civil cases.

These reforms, significantly, were pushed through by Governors superior in rank to Captain, for the advent of the American War of Independence had suddenly enhanced Newfoundland's strategic significance, making the appointment of a Rear- or Vice-Admiral, with the capacity for taking decisive action, an urgent priority. Most of them, expecting no further advancement, did not fear provoking controversy or even making enemies in their dealings with the recalcitrant Newfoundlanders, and outfaced any opposition to reform. In the meantime, the harsh discipline of the Royal Navy was inflicted on those convicted of misdemeanours or, alternatively, fines of ten to twenty pounds (equivalent to a season's wages) were imposed for relatively trifling offences by justices and their clerks who were entitled to keep a share of the monies thus raised. It is hardly surprising if habits developed among the population of winking at infractions of the law and of refusing to co-operate with such an oppressive system of justice.

The stress imposed on the landless Newfoundlander by the daily struggle for subsistence, and the virtual absence of the normal institutions of civilized life, was exacerbated by the threat of armed conflict. Once the British, galvanized by William Pitt the Elder, decided to challenge the might of France in India, the Caribbean, and in Canada, then Newfoundland became a small, but nonetheless significant, piece in an intricate game which was played out in earnest across the globe. The crisis came in the 1760s, towards the end of the Seven Years' War, when Spain secretly made common cause with France in the hope of re-establishing her ancient fishing rights off Newfoundland. Pitt, the Prime Minister, perceived the danger in this, but failed to persuade his colleagues in the administration of the expediency of a pre-emptive strike against the Spanish fleet, and so felt compelled to resign. His fears proved justified for, in 1762, when Spain had finally declared her hand, so that Britain's navy had to contend with both rival powers simultaneously, Newfoundland, whose defences had been neglected, fell easily to a French squadron. The inhabitants of the principal harbours had to suffer all the indignities of occupation for three months until a force was mustered to evict the French garrison — an operation which was effected with characteristic vigour.

During the preliminaries to the treaty which formally marked an end to hostilities, the British negotiators were amazed by an unsolicited offer on the part of France to give up the whole of Canada in exchange for the two islands of St Pierre and Miquelon which lie off Newfoundland's southern coast. But France, since losing Quebec in 1759, placed a higher value on a toe-hold in those waters than on the other Canadian possessions which might have remained to her. Not only did she thereby enhance her capacity to remain a first-class naval power, but also won the right to fish along the whole length of the coast from Cape St John in the east right round to Cape Ray in the south west, while the British retained exclusive access to the Great Bank. This, of course, gave to Great Britain the lion's share of the export trade which by the end of the eighteenth century amounted to 500 000 quintals (i.e. hundredweight) of cod per annum. The Government in London congratulated itself on the successful outcome of these negotiations, but a Dr Gardner of Boston, who settled in Newfoundland after the American Revolution, noted that the Great Bank could be fished only from April to November, while the French had access to inshore fishing virtually all year round. 'This,' he commented, 'is a full proof we are not equal to that subtle, artful people in the cabinet, although we are in the field'.[1]

The treaty, agreed at St Germain in 1763, and ratified in London on 10 February 1764, (and confirmed in 1783 in the armistice at the end of the American War of Independence), included provision for the liberty of conscience, 'so far as the laws of Great Britain allow', for the Catholic population of what had hitherto been French North American colonies. For a time, Quebec was linked to Newfoundland for administrative purposes, with the result that the land-hungry sons of Ireland's poor farmers, taking advantage of the unexpected removal of impediments to religious association and practice which was provided for by the treaty, began to emigrate in significant numbers. This was coupled with an increased presence of Catholic clergy — much to the chagrin of Church of England missionaries of the Society for the Propagation of the Gospel. The latter had also to contend with the increased activities of

Methodist missionaries who linked their preaching to questions of social justice — or, we might rather say, *in*justice — which had been neglected for so long.

The correspondence of the Society missionaries paints a depressing, one might even say tragic, picture of life in Newfoundland in the late eighteenth century. By following the career of just one such clergyman we can appreciate the conditions which led inevitably to a hand-to-mouth existence for considerable numbers of the population there. The Revd James Balfour arrived in Trinity Bay in 1764 where he remained for ten years. In his first letter to the Society he commented that the people were good natured but that 'they may be led but not drove'. A month later, in October, we find him complaining of having few parishioners on a Sunday on account of some 'drunken villains' having burnt down the woods near the town, forcing women and children to go far afield in search of firewood. Although most of the residents were very poor, nonetheless he found it very difficult to find lodging in case his presence 'should check some favourite vice'. Marriage, he noted, was often dispensed with and there was a general 'laxity of morals'. During 1765–6 his parishioners built him a good house and everything seemed settled and tranquil. However, by 1767 he was reporting much 'feuding' among the people — whether over local or political matters or religious doctrine is unclear — and, as trade had declined, people were not paying their pew rents, with the result that the church building was on the point of falling down.

The presence at this time of what he called 'lawless Irish immigrants' who 'rob and house-break' is to be noted; indeed complaints about 'Irish malcontents' punctuate his letters during succeeding years, as do observations on the wretched plight of the poor. The winter of 1768–9 was particularly harsh, with a severe frost lasting until early April, and the population suffered 'an epidemical disorder of a swimming in the head or giddiness'. In late 1771 Mr Balfour complained that the people of Bonavista were agitating for a minister (i.e. a dissenter from Anglican doctrine), which he construed as wickedness and faithlessness on their part. He struggled on until 1774 and then asked for some reward in view of his long service, commenting 'believe me these are uncouth regions here indeed for a man to spend his short life in'.

The following year James Balfour was given a new living not far round the coast at Harbour Grace where he found the property owners 'strictly loyal', but noted a different temper among the bulk of the population who, 'were there numerous and enterprising men to head them, would exactly resemble the Americans on the continent'. We may infer that he was, to use an old phrase, 'stiff for the King and the Church of England', for in December 1776 he complained that the turn of events in the revolted colonies in North America had made his parishioners 'truculent' towards him. In consequence he had to enlist the help of the Governor, Rear-Admiral Montague, to obtain a house.

By 1779, the economic effects of the American War of Independence had forced some of the population to emigrate to New England or to return to England. Visiting ships had been taken as prizes by American privateers (pirates in effect) — a fact which must have discouraged others of whatever nation who would have bought and sold in Newfoundland. 'In short,' he wrote, 'there is a raging famine, nakedness and sickness in these parts' which was made the more intolerable by the incidence of house-breaking by 'desperadoes'. The Society in London must have been alarmed to receive such correspondence from one of their missionaries who, in this instance, seems to have been at his wits' end, concluding his letter with the pathetic question, 'What shall I do?'

Balfour continued for a number of years but in the end seems to have despaired even of the support of the representatives of the crown in Newfoundland. Rear-Admiral Edwards, who was Governor of the Island in the early 1780s, on receiving Balfour's application for redress for having been insulted while preaching at Carbonear, agreed only to reprimand those concerned and 'seemed to think that the chapel at Carbonear was private property and the people could do what they chose in it'. Feeling badly let down, James Balfour asked to be recalled after more than twenty years' service and to be given a living in England.

Fear of French intentions led to a corresponding anxiety about the presence of so many Irish immigrants who, being both Catholic and justifiably resentful of the economic despoliation of their homeland, were considered by the powers that be as potential allies of Britain's historic rival in the region. One immigrant, a certain George Tobin, went so far as to hoist a standard displaying Irish colours, for which he was summarily fined the enormous sum of ten pounds. When hostilities were renewed with the onset of the French Revolution, the Catholic Archbishop was moved to issue a strongly worded statement in favour of loyalty to the British Crown as against the specious arguments of the revolutionaries. A heartily grateful Governor ensured that he received a handsome pension from the British Government for this timely intervention.

Towards the end of the eighteenth century the population of Newfoundland began to increase rapidly, for in spite of privation in the winter and the exorbitant prices charged by merchants who had a monopoly of certain commodities, the promise of good wages in the fishing season still offered the prospect of escape from poverty at home. The frustration and disappointment which must have been felt by many were summarized by Dr Gardner, whose pertinent observations on the negotiating skills of the French have already been cited in this chapter:

> The common fishermen...are hired from ten to thirteen pounds for two summers and one winter, but as the merchants cannot employ so many of them [as] ten thousand in the winter, as they can during the fishing seasons, of course they

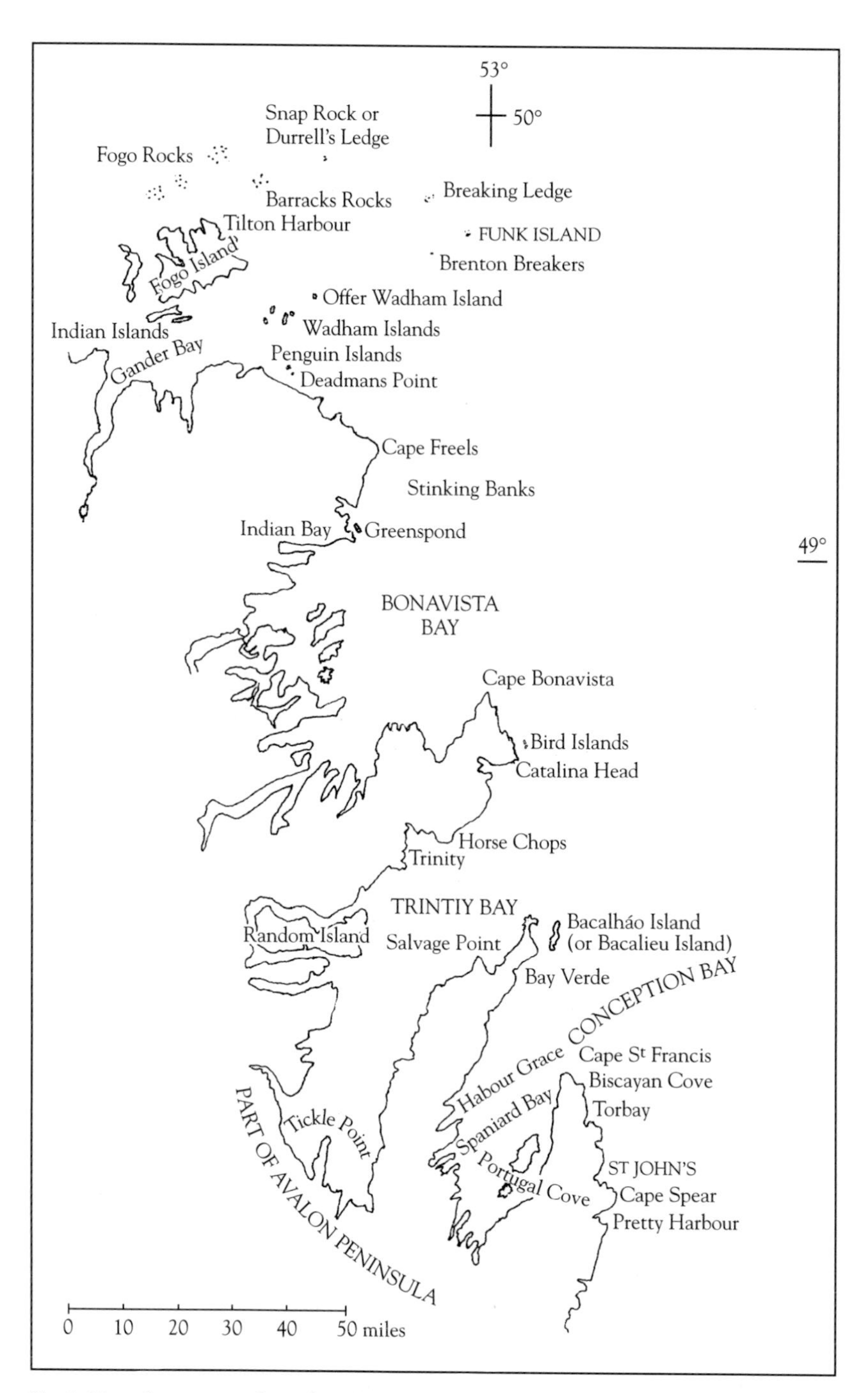

Fig. 1 Map of east coast of Newfoundland from Fogo Island to St John's.

> must either stay there, and starve, or return home, without any hope of ever bettering their circumstances. — It likewise appears impossible that the servant can have saved so much from his wages as will support him in England, or Ireland, during the winter season, more especially if he has not secured the return of his passage in his original contract, which is too often omitted.

It might be thought that there were enough people living in Newfoundland by Dr Gardner's time to make the hiring of more fishermen from England and Ireland unnecessary. A number of coastal communities, however, were restricted in their operations by the successive treaties between Great Britain and France. The little community known as Greenspond (or Green's Pond), 120 miles north of St John's, had, since the Treaty of Utrecht in 1713, been on that part of the coast whose adjacent waters were reserved for the French fishing fleets. A decline in the number of accessible cod in those waters during the early eighteenth century led to economic distress, but for a decade after 1730 this situation was reversed so that Greenspond actually saw an influx of immigrants. However, the decline set in again after 1740 so that the population which had reached some three hundred souls had to look north, to as far as the Labrador coast, for their livelihood. About 60 miles north-north-east of Greenspond lies Funk Island, and undoubtedly some of the crews soon learned that a few days' work killing sea birds there was equivalent to many weeks spent at sea fishing for cod far from home. That the wages were, by local standards, good explains, in part at least, why in time Greenspond earned the nickname 'the Sodom of the North'.

With a rising population and the acute distress of the poor it soon became clear, simply by virtue of the laws of supply and demand, that good money was to be made from eggs and birds. Able-Seaman Aaron Thomas, a crew member of HMS *Boston*, learnt about the Great Auk's breeding grounds when his ship put into St John's in 1794. In his journal, which was written in the form of an extended letter to a friend, Thomas gave a colourful account of what he had been told of the activities there in the breeding season:

> If you go to the Funks for Eggs, to be certain of getting them fresh, you pursue the following Rule:- you drive, knock and shove the poor Penguins in Heaps ! You then scrape all the Eggs in Tumps in the same manner you would a Heap of Apples in an Orchard in Herefordshire. Numbers of these Eggs, from being dropped some time, are stale and useless, but you having cleared a space of ground the circumference of which is equal to the quantity of Eggs you want, you retire for a day or two behind some Rock at the end of which time you will find plenty of Eggs — fresh for certain ! — on the place where you had before cleared.[2]

Aaron Thomas was given an idea of the profits which could be had from these expeditions by his informant: in two trips, with one other person, this friend had secured 'as many eggs as sold at St John's for thirty pounds'. When salted, these eggs, selling for a few pence a dozen, were a good standby for the poor during the winter, and indeed the birds themselves were a cheap substitute for salted pork.[3] There is, of course, a significant difference in degree between the hunter taking eggs or birds for his own immediate use or that of his family, and taking them as part of a speculative enterprise; accordingly, it is likely that all of the Alcine species, including the Great Auk, could have continued to withstand occasional raids of the kind described, as they had done ever since Jacques Cartier first discovered the *Isle des Oyseaux* in the mid-1530s. However, as the fowlers were, by the eighteenth century, in competition with each other, there was no incentive to ensure that sufficient birds were left to gather during the following season. Anyone who failed to kill and grab as much as he could carry away knew that those who came after him would not hesitate to fill their boats with whatever was left. It is easy to see how different things would have been if exploitation of the birds on Funk Island had been organized co-operatively by those who made annual visits there.

Egg gathering after the manner described by Aaron Thomas was to have a noticeable impact during the nineteenth century on the populations of all those sea birds whose colonies were accessible from the land. Systematic plunder of this kind, particularly if the reproductive rate of a species was low, could quite easily have led to the extermination of any, such as the Great Auk, with only one or two suitable breeding localities. It is very doubtful if more than a few at most of the Great Auks of Funk Island would have succeeded in bringing off their young in the last decades of the eighteenth century and, given the scale of commercial egging at this time, there is little doubt that the continued survival of the species would have been threatened even if the adult birds themselves had been left unmolested.

The decision to appoint Governors of above the rank of Captain or Commodore from 1776 onwards resulted, as we have said, in some long-needed social amelioration in Newfoundland. Lewis-Amadeus Anspach, the first writer of note to compile a comprehensive history of the island, had little time for those affected by the spirit of republicanism which was already abroad, but, rather, had the highest opinion of the senior naval officers who enjoyed plenipotentiary powers. Of Vice-Admiral Waldegrave, Governor of Newfoundland 1797–1800, he wrote in characteristic vein:

> His attention to the interests of Religion, and to the due administration of Justice, was an insurmountable obstacle to his acquiring any popularity in St John's where Paine's *Age of Reason* and *The Rights of Man* had poisoned the public mind and had more weight than either the Bible or the Acts of Parliament.

Of Rear-Admiral Elliot's term as Governor from 1786–9, Anspach concluded that the difficulties encountered in establishing order and the administration of justice 'were such as must have discouraged and disgusted a man of less judgement, prudence, and perseverance'.

There were many vested interests to be overcome before a Court of Common Pleas was established at the end of Governor Elliot's period of office. The West of England merchants and shipowners had always preferred the previous inefficient system of justice which they could either use to their advantage or obstruct as it suited them. 'They opposed,' wrote Chief Justice John Reeves in 1793, 'every attempt at introducing order and government into that place.' The fishing-admirals, supported by their financial backers, had even had the temerity to argue that the only authority on the island had been the King in Council, not Parliament itself, and by employing such constitutional arguments tried to assert their traditional rights as against the new institutions of civil government. Popular representations were also made — orchestrated one suspects — in support of the fishery in Newfoundland continuing to be 'free', repeating the refrain that a fishery carried on from England 'as the western merchants carried it on was the old and true policy for Newfoundland'.

Between 1780 and 1810 steps were taken, again long overdue, not only to put the administration of justice on a proper footing but also to effect a reconciliation with the surviving Beothuk Red Indians. An earlier attempt, led by Captain Cartwright in 1768, had failed to make contact with these indigenous people who were, by then, terrified of Europeans. The proclamation read by Governor Montague on 6 May 1776 is remarkable testimony to the new style of government which Newfoundlanders found themselves having to grow accustomed to. The reports state that the proclamation commenced with the information that:

> ...it had been represented to his Majesty that his subjects residing in Newfoundland, instead of cultivating such friendly intercourse with the native savages inhabiting the island as might be to their mutual benefit, treated the said savages with the greatest inhumanity, and, in some instances had destroyed them without the least provocation; that it was his Majesty's royal will and pleasure that he, the Governor, should express his Majesty's abhorrence of such inhuman barbarity.

It is difficult to believe that it is simply a coincidence that this proclamation was made only two months before the American Declaration of Independence. Whatever the merits of the political arguments adduced in favour of a separation from the mother country, and however insensitive Great Britain's treatment of her colonies on the American continent had been, it is an undeniable fact that part at least of what motivated the most influential supporters of a breach was the prospect of being able to disregard treaties with North American Indian tribes in respect of land ownership. The British Government — and we should recognize that the royal prerogative by this time was exercised collectively by the Prime Minister and other members of the Privy Council, not by the King in person — was fully aware of these developments and had become acutely concerned that its honour would be impugned and the reliability of

Fig. 2 Engraving of Captain George Cartwright, from his Labrador *Journal* (1798).

international law undermined, if they permitted these North American treaties to be regarded as mere scraps of paper to be discarded at will. Consequently, Red Indian affairs were at the forefront of the minds of ministers throughout the 1770s and hence the condition of the Beothuks came under scrutiny. Governor Montague's proclamation, we are told:

> ...concluded by strictly enjoining and requiring all his Majesty's subjects to live in amity with the said savages; and all officers and magistrates to use their utmost diligence to discover and apprehend all the persons who might be guilty of murdering the said Indians so that such offenders might be sent over to England for trial.

Other steps taken by the Admiral-Governors of the era are, perhaps, even more remarkable than their earnest attempts to save the indigenous Beothuks. Admiral Pole, in 1805, received petitioners who sought redress against a merchant class which was widely regarded as unscrupulous and tyrannous: '[T]he said merchants,' he was informed:

> ...arrogate to themselves a power not warranted by any law, in selling to us every article of theirs at any prices they think fit and taking from your petitioners the produce of the whole year at whatsoever price they think fit to give...In short let it suffice to inform your Excellency that they take it upon themselves to price their own goods and ours also as they think most convenient to them.

Remarkably — to a later generation at least — Admiral Pole acted as champion of the poor in St John's by having recourse to the crude but effective expedient of pegging prices at the previous year's level in the event of sustainable accusations of unfair practice.

The first Governor of the nineteenth century, Vice-Admiral James Gambier, was a remarkable, even revolutionary, educational reformer. Recognizing that the endemic problems associated with illiteracy and a lack of moral formation at a youthful age would persist in Newfoundland while the few schools which existed were either fee-paying or relied exclusively on voluntary contribution, he insisted that universal elementary education should be paid for by everyone according to their means. He immediately set an example by contributing forty pounds — a very considerable sum at that time — from his own pocket.

That the naval administration of Newfoundland during these years was also responsible for the introduction, if only by proclamation, of the first legislation for the protection of sea birds known in the Western World, is another manifestation of the humane policies of what was a remarkably innovative and broad-minded succession of Governors whose profession and rank might lead us to suppose that the opposite would have been the case. All this, moreover, was achieved against a background of tension and uncertainty caused by the limitless ambition of Napoleon Bonaparte.

The evident distaste felt by Anspach, and those of a similar political conviction, for the radicalism of the followers of Tom Paine[4] was not based entirely on prejudice. In 1797, the year in which the crews of the British fleet which lay between England and invasion had mutinied on account of their conditions of service, a similar disturbance caused anxiety at St John's — though it was confined to but one ship, *The Latona.* Despite the fact that the men held the officers hostage, Governor Waldegrave vowed that he would not hesitate to fire on them from the shore 'and burn [the ship] with red-hot shot in case you drive me by your mutinous behaviour to that extremity'. Fortunately the crisis was resolved without recourse to such drastic measures. In February 1800, a conspiracy was hatched by a group of United Irishmen in the Royal Newfoundland Regiment, supporters of the revolutionary Wolf Tone, to take Fort Townshend. It was discovered only by chance by the military authorities at the last moment. As a result, five men were summarily hanged in the powder-house of the fort, seven were subsequently shot, and others were sent in irons to Halifax, Nova Scotia.

The 1790s also saw an outbreak of smallpox on the island, presumably inadvertently brought there by a newcomer, and, in view of the absence of medical screening, it is not surprising that there were other such epidemics in 1800 and 1813. Additionally, some sections of the population still suffered terrible destitution in the winter. In order to discourage anyone from the British Isles from coming for the fishing season — unless he had enough money saved to enable him to survive the winter — poor relief was stopped in 1802. That this was not the end of social responsibility as far as persons in authority were concerned is borne out by a remarkable testimonial of thanks given by the inhabitants of St John's to a Captain Buchan[5] who, in the depths of a particularly severe winter when:

> ...the labouring class of people were no longer able under their half-famished condition to support the usual winter's toil...when the purses of the inhabitants were drained by the constant contributions and when even donations in money to the poor were unavailing to relieve them since provisions were not to be bought at any price

put his crew on reduced rations and made over to the public whatever he could to relieve 'their alarming and terrible wants'.

As if anything else were needed to render life precarious for everyone, but especially the poor, St John's was severely damaged by fire on three occasions between 1790 and 1820. Moreover, two of the fires occurred in successive years, which led to a widespread belief that they had been started deliberately — although no evidence to substantiate this ever came to light. Anspach gave a succinct description of the fire of 1816, which in many respects encapsulates the rigours of life as the eighteenth century gave way to the next:

> The conduct of the seamen from the King's ships, and of troops from the garrisons, as well of the respectable parts of the inhabitants, was represented on this critical occasion, as beyond praise, while the bulk of the lower orders stood, with their arms folded, surveying this disastrous scene with an apathy disgraceful to the human character and appearing to have no object but pillage. The aggregate pecuniary loss occasioned by this conflagration was estimated at upwards of a hundred thousand pounds sterling; and about fifteen hundred persons were driven to seek new abodes in February, the most inclement month of a Newfoundland winter. The distresses of these unfortunate sufferers were considerably aggravated by the depredations committed by the populace upon the property snatched from the flames.

The fire in the year following, which destroyed the wharves when barrels of tar were ignited, was considered to be much more severe in terms of financial loss, the final estimate of which being as much as half a million pounds. Many must have been thankful, however, that the destruction was confined to the harbour and did not drive them out of their newly reconstructed homes.

Notes

1 Gardner's opinion makes an interesting comparison with the views of Pitt who, in the House of Commons, opposed the terms of the treaty: 'France is chiefly, if not solely, to be dreaded by us in the light of a maritime and commercial power: and therefore by restoring to her all the valuable West India

islands, and by our concessions in the Newfoundland fishery, we have given her the means of recovering her prodigious losses and of becoming once more formidable to us at sea.'

2 The natural inference from this is that the Great Auks would lay a second time if they lost their first eggs — something which was expressly denied by Martin Martin in his account of the species on St Kilda.

3 The authority for this tradition, repeated by a number of subsequent writers on Newfoundland, is Captain Cartwright's *Journal.* The context suggests that the meat of Guillemots (Murres), Razorbills, and Puffins, as well as that of the Great Auks, was sold for this purpose: 'The birds which the people bring from thence they salt and eat in lieu of salted pork.'

4 Tom Paine, 1737–1809, an eloquent advocate of rationalism and republicanism, was a notable political propagandist: at the time of the Declaration of Independence his *Common Sense* was instrumental in convincing many waverers that separation from Great Britain was the only way for the colonies to reach their full potential.

5 Captain Buchan had, in addition, been the leader of the party sent in 1810 to try, with only limited success, to achieve a reconciliation with the Beothuk Indians.

CHAPTER 10

Mercenary and cruel

In the summer of 1853, a boatman named William Stabb, by then aged about seventy and living in semi-retirement in the popular holiday resort of Torquay in south Devon, recalled his younger days in Newfoundland as he chatted to two keen naturalists, Alfred and Edward[1] Newton. The former of these two brothers had already, at the age of twenty-four, acquired the reputation of being one of Britain's foremost ornithologists and became in time the leading authority on the Great Auk in the English-speaking world. Moreover, for forty years from 1866, as Professor of Zoology and Comparative Anatomy in the University of Cambridge, he was able to exercise considerable influence both as expert witness and member of the committee established to lobby for statutory protection for wild birds — once public opinion became receptive to such a law. In addition, he was a founder member of the British Ornithologists' Union and was in large part responsible for establishing its journal the *Ibis* which is still published today. Alfred Newton was, in short, one of the most influential ornithologists of the latter half of the nineteenth century — a period which saw a tenfold increase in the volume of knowledge of this branch of natural history as the avifaunas of previously neglected parts of the world yielded up their treasures to European collectors.

William Stabb informed his listeners that he used to ' follow the Newfoundland cod fishery and that he had seen penguins off that coast'. He added that they used to 'resort by hundreds to some islands there to breed, but were destroyed for their feathers, being driven up in a corner by people in boats'. In reflecting on this a decade later, Alfred Newton inferred: 'This practice, however, must have nearly or altogether ceased in his time; for he stated that he had never seen but two or three birds himself, and never a dead one.' This accords with what was attested by others — that by the end of the eighteenth century, the population of Great Auks on the Great Bank and the seas adjacent, which had numbered tens (if not hundreds) of thousands in the 1530s when Jacques Cartier was prospecting for the fabled North West Passage, had been reduced to a pitiable remnant.

During the two hundred or more years prior to 1800 in which the populations of sea birds were annually culled for provision against the winter, the Great Auk

Fig. 1 Central flat-top colony of Common Murres (Common Guillemots) at Funk Island, Newfoundland from the ground (11 July 1975). This view shows the likely area where the major concentration of breeding Great Auks occurred. (Photograph by D.N. Nettleship)

was particularly targeted not only on account of its size but also because, as the Newfoundland ornithologist, the late Lesley Tuck pointed out, the birds could be killed without recourse to firearms, the ammunition for which was too expensive for liberal use. As it has been suggested, progressive decimation over the years does not, on its own, account for the annihilation of the Great Auk. Instead we should regard it as an important contributory factor which had the predictable result that once the sea bird populations came to be regarded as a source of profit and not merely of sustenance, the remaining Great Auks at a limited number of breeding sites could be extirpated only too rapidly if exploitation went unregulated.

As Funk Island, which was the last substantial refuge of the species, lies 170 miles north west of St John's, we should not imagine that the people of the island's capital regularly took part in the traditional raiding parties every June and July. This was the preserve of the inhabitants of Newfoundland's northern shore, from Cape Bonavista in the east to Twillingate in the west, a sparsely populated region where we find such aptly named communities as Seldom-Come-By. In particular, the people of Fogo Island were dependent on the Funks for survival from November to March, for this stretch of the coast was part of the extensive area of fishing rights, known as the French Shore, which as related in the preceding chapter, was accorded to France under the treaty ending the Seven Years' War.

Captain George Cartwright, who knew the coasts of Labrador and Newfoundland better than any of his contemporaries, saw at once the gravity of a situation which had graduated from piecemeal exploitation to sustained rapacity. On 25 July 1785, Cartwright spent the day in the company of the officers and men of a British survey ship at St John's and during the day a 'boat came in from Funk Island laden with birds, chiefly penguins'. Of the island itself he wrote:

> Innumerable flocks of sea fowl breed upon it every summer which are of great service to the poor inhabitants of Fogo; who make voyages there to load with birds and eggs... But it has been customary of late years for several crews of men to live all the summer on that island, for the sole purpose of killing birds for the sake of their feathers; the destruction which they have made is incredible. If a stop is not soon put to that practice the whole breed will be diminished to almost nothing, particularly the penguins: for this is now the only island they have left to breed upon.

The date is not insignificant for it coincides with the acute economic privation which resulted from the American War of Independence and the acts of piracy associated with it. An anonymous letter, signed simply 'A Fisherman', which first appeared in a New Jersey newspaper, the *Salem Register*, and which was reprinted in *The Gloucester Telegraph* (another New Jersey publication) in early August 1839, made the same point from a different perspective:

> All the mackerel men who arrive report the scarcity of this fish[2] and at the same time I notice an increase in taking them with nets at Cape Cod and other places.
>
> If this speculation is to go on without being checked or regulated by the Government, will not these fish be as scarce on the coast as penguins are, which were so plenty before the revolutionary war that our fishermen could take them with their gaffs ? But during the war some mercenary and cruel individuals used to visit the islands on the eastern coast which were the haunts of these birds for breeding, and take them for the sake of the fat [*sic*],[3] which they procured and then let the birds go. This proceeding destroyed the whole race.

We find evidence of the profits envisaged in this trade in an account of the French Shore written in 1847 by a certain Captain Loch to the Governor of Newfoundland, the Earl of Dundonald, (better known to history as the brilliant but headstrong Lord Cochrane):

> At noon we passed the Funk Islands within a mile, leaving it on the port hand. Parties repair thither in spring and autumn, to collect eggs and feathers. At one

> time a very considerable profit could be gained by this trade, but lately, owing to the war of extermination, that has been waged against the flights of Puffins, Gannets, Divers,[4] Gulls, Eidar Duck, Cormorants, etc., etc., it has greatly diminished. One vessel of 25 tons, is said, once to have cleared £200 currency on a single trip to Halifax.

Halifax was an important naval base on Nova Scotia; commencing in 1748, it took eleven years to build as the rival to the French citadel at Louisbourg on the Cape Breton peninsula.[5] For forty-four of the first sixty-six years of its existence, Halifax was maintained in a state of war with either France or the United States of America. It is likely that only those prepared to put their lives at risk engaged in commercial activity beneath its battlements. Among these 'freebooters' we include the blockade runners, ready to incur the displeasure of governmental authorities by trading with nominally hostile sea port towns. These men knew that however much they paid to the Newfoundlanders, they could still make a substantial profit when they supplied feathers to the wholesale markets in Maine, New Hampshire, and Massachusetts.

In view of this unbridled economic activity, it is easy to sympathize with the following outburst from F.A. Lucas who visited the waters about Funk Island in 1887, ostensibly to enquire into the mackerel fishery, but with the ulterior objective of collecting Great Auk bones for the Smithsonian Institution where he worked as assistant curator:

> It is now about fifty years since the Great Auk succumbed to the incessant persecution of man, disappeared almost simultaneously from the shores of Europe and North America, and became extinct. Found along the coast of Newfoundland by the early explorers, the countless myriads of this flightless fowl had been hunted to the death with the murderous instincts and disregard for the morrow so characteristic of the white race.

On reflection, the argument regarding racial characteristics, which, in the nineteenth century was regarded as intellectually respectable, is a poor substitute for an analysis of the economic and cultural conditioning which has prevailed in the Anglo-Saxon world since the publication of Adam Smith's *Wealth of Nations.* In fact, as we have seen in the *Journal* of Captain Cartwright, there were voices raised in opposition to what was then happening. As ever in Newfoundland, the stand-off between Government and governed, exacerbated by legislation simply handed down by lawful authority with scant regard to public opinion, ensured that, in the absence of robust enforcement, the King's writ hardly ran at all outside the island's capital. In discussing Guillemots — or 'Baccaloo birds' as they were known locally after one of their principal breeding sites, Baccalao Island — Lewis-Amadeus Anspach in his *History* of the island observed that they:

> ...have ever been considered as of sufficient importance to mariners, particularly in foggy weather, by giving them notice of their approach to the coast, even as far

> as the Banks, as to deserve the special protection of Government against the attempts of birds' and eggs' hunters. Notwithstanding the Proclamations issued, from time to time, by the Governors of Newfoundland for that particular object, it has not infrequently happened that, tempted by the vast profit produced by the sale of those birds, of their eggs, and of their feathers, and regardless of the extreme dangers which attend the attempt, some daring individuals contrived...to make a general sweep of the eggs as well as of the birds themselves.

The first such proclamation was issued by Rear-Admiral John Elliot in 1786 and its proximity to the date of Captain Cartwright's reflections committed to paper in July the previous year, makes it probable that none other than Cartwright himself had made representations, in quarters where he was held in high regard, to ensure that a stop was put to the annual slaughter on the Funks. Cartwright was, as we have seen, particularly concerned about the fate of the Great Auks there, and this species was a more accurate indicator of 'the proper soundings' than even the local Murres (Guillemots), as we may read in the 1728 edition, and in all the subsequent editions, of the standard reference work for coastal navigation, *The English Pilot*:

> There is another thing to be taken notice of in treating of this coast, that you may know this by the great quantities of fowls above the Banks, viz. shearwaters, willocks [guillemots], noddies [fulmars], gulls, penguins etc. without making any exception; which is a mistake for I have seen all these fowls an hundred leagues off this bank, the penguins excepted. It is true that all these fowls are seen there in great quantities, but none are to be so much minded as the penguins for these never go without the Bank as others do, for they are always on it or within it, several of them together, sometimes more, sometimes less, but never less than two together.

So vulnerable was the Great Auk by the last decades of the eighteenth century that it wanted only a few people acting on the principle that 'if you live by the law, you don't *live* at all' to reduce such a proclamation as the one read by Governor Elliot in 1786 to little more than moral exhortation.

Faced with revolution in both America and, in more bloody guise, in France, the land-owning gentlemen who controlled public life in Great Britain reacted to seething discontent when it manifested itself at home by active repression and the obstruction of demands for radical reform. Newfoundland, in consequence, was not accorded a legislative assembly and colonial status until 1824, and in the preceding decades, the naval Governors of the island knew it was incumbent upon them to 'run a tight ship' as well as to address abuses in the administration of justice. Not infrequently the deliberate flouting of the law continued to be met, in the time-honoured way, with a public flogging. To the civilian population this was a 'cruel and unusual' punishment which often increased their sympathy for the convicted party. Consequently those who were tempted by the prospect of the 'vast profit' described by Anspach could be fairly sure that anyone they met in the vicinity of the breeding grounds of the Great Auk would be very unlikely to report them. The only significant risk was of falling in with a naval vessel, but it was evidently a risk well worth taking. Aaron Thomas whose *Journal*, written in 1794, has been referred to in the previous chapter, passed on what he had heard of the trade:

> [S]kinning and taking the eggs from the Funks is now prohibit'd and they are allow'd to take the Birds only for Bait to catch Fish with. The Funks being at a distance from the land, are so uninviting and desolate that they are seldom viseted [*sic*] unless by Pirates and Robbers to steal the Feathers and Eggs. About three years ago some fellows were detected in this kind of plunder. They were brought to St John's and flogged at a cart's tail. But I am told there is quantitys of feathers purloined from these islands every year.

It would be of interest to know whether the 'Pirates and Robbers' whom Thomas had in mind were local people or whether, perhaps, some of the entrepreneurial characters who traded from Halifax had discovered the location of Funk Island and were securing supplies of feathers for themselves. Because these activities were strictly illegal after 1786, there is, as might be expected, little accurate information about what actually went on there. However, in 1876 the American ornithologist Joel A. Allen received from an expert on the Newfoundland sealing industry, Michael Carroll of Bonavista, information suggesting that the Great Auk was remorselessly pursued even as late as the 1820s. Allen summarized what he had heard in *The American Naturalist*:

> In early life he was often a visitor to [the] islands, and a witness of what he here describes. He says these birds were formerly very numerous on the Funk Islands, and forty five to fifty years ago were hunted for their feathers, soon after which time they were wholly exterminated. As the auks could not fly the fishermen would surround them in small boats and drive them ashore into pounds previously constructed of stones. The birds were then easily killed, and their feathers removed by immersing the birds in scalding water which was ready at hand in large kettles [i.e. cauldrons] set for the purpose.[6] The bodies were used as fuel for boiling the water. The wholesale slaughter, as may well be supposed, exterminated these helpless birds, none having been seen there, according to Mr Carroll, for more than thirty years, and he expresses great doubt in respect to the existence of the species now anywhere about the islands of Newfoundland or Labrador.

Something similar had appeared in print in London forty years earlier, but until now has gone unnoticed. In 1836 the Revd Edward Wix published *Six Months of a Newfoundland Missionary's Journal* which contains the following laconic entry for 15 April 1835:

> I looked today over the whaling establishment of Messrs. Hunt and Newman...The refuse pieces of whale, which are left in the boiler, after the oil is

Fig. 2 Feather hunters approaching Funk Island, *c.*1780.

> extracted, furnish, I am informed, all the fuel which is required for heating the coppers. This recalls to my recollection the fact, that the early settlers on this island used to make fires with the piles of the carcases of fat penguins, a bird which used then to be very common, but is now extinct, or has left the island. They were most cruelly treated while they abounded in the island, being often plucked for their feathers and then turned loose to perish, or burnt in piles as above described.

Because Mr Wix did not arrive in Newfoundland until 1827, he had no first-hand knowledge of these misdeeds and in ascribing them to the early settlers he displayed a lack of awareness about when the events took place. The description itself, however, bears all the hallmarks of being unembellished hearsay and contains nothing to suggest that Wix's informant did not know what he was talking about. Further evidence of deliberate cruelty is to be found in the *Journal* of Aaron Thomas, but these accounts appear to have been unknown[7] to the compilers of the first nineteenth-century histories of the Great Auk. Thomas emphasized that an acquaintance in St John's (who, as recalled from Chapter 9, in two trips to the Funks took as many eggs as sold for thirty pounds 'when this kind of Traffick was lawful') was someone whose veracity he could depend upon:[8]

> If you come for their feathers you do not give yourself the trouble of killing them, but lay hold of one and pluck off the best of the Feathers. Then you turn the poor Penguin adrift, with his skin half naked and torn off, to perish at his leasure [*sic*]. This is not a very humane method, but it is the common Practize...
>
> While you abide on this Island you are in the constant practize of horrid crueltys for you not only Skin them Alive, but you burn them Alive also, to cook their bodies with. You take a kettle with you into which you put a Penguin or two, you kindle a fire under it, and this fire is absolutely made of the unfortunate Penguins themselves. Their Bodys being oily soon produce a Flame; there is no wood on the island.

There can be few other species among those which have been driven to extinction of which it could be said, in the words of Symington Grieve, a late nineteenth-century writer on the Great Auk, that some scenes in their life history 'told with pathos might bring tears of sympathy from hearts of stone'.

It may be asked whether there existed a unique combination of circumstances in eighteenth-century Newfoundland which account for the kind of excess just related. Tempting as it may be to excuse what happened on account of the conditions which existed in one particular dependency of the British Crown at a certain stage in its development, the truth is that there are uncomfortable parallels to be drawn between the social conditions on the island and those in Great Britain at that time. Probably the most important of these was the lack of any formal education for the great mass of the poor — an issue addressed earlier in Newfoundland than it was in England where it took the social explosion which preceded the Great Reform Act of 1832 to act as a catalyst.

Both the unhappy consequences of leaving a large section of the population without an elementary education, and also the beneficial effects of persisting in an enlightened policy, can be observed in the following remarkable passage from *Macmillan's Magazine*, published in London in 1868:

> When, in 1836, a few gentlemen began to stock the lake in St James's Park with waterfowl, the rough frequenters of the park — men and women, as well as children — startled at the unaccustomed sight of the birds, destroyed them in immense numbers, and if any one more rare and curious than the rest appeared on the water, he was immediately made a special 'cockshy' for stones and killed. A male Smew (*Mergus albellus*), the first that was known to the memory of man to come alive into the London market, was bought one day in 1837, and turned out upon the lake. He went rushing up and down, now diving and erecting his wings, till he attracted a great crowd; alas! they perseveringly pelted him with stones till he died within an hour of his first appearance on that watery stage. The Ornithological Society was formed, and once a week the committee received lists of birds killed and wounded by missiles during the preceding se'nnight. The losses were so numerous, and the expense of replacing them so difficult to be met that it was seriously debated whether the Society should not give up its enterprise, on account of what seemed to be the incorrigible habits of cruelty and mischief of the people. Happily it was resolved to persevere, in the hope that, after a while, the public would become interested in the birds and no longer persecute and annoy them. Everyone knows that this hope has been completely realised. The Ornithological Society has stocked all the waters in the parks with waterfowl; not only St James's, but Hyde Park, the Regent's, Victoria and Battersea Parks; and nothing is more rare than any injury wilfully done to the birds.

As Sir Richard Bonnycastle stated in his *Newfoundland in 1842*, 'it was the ruthless trade in its eggs and skin' (by which we can assume he meant the feathers) that led to the disappearance of the Penguin of Newfoundland.

Success in limiting the exploitation of sea birds only began to be achieved with the imposition of swingeing fines which were introduced, ironically enough, just as the few remaining Great Auks were disappearing from their last retreat in European waters. Later in the nineteenth century, the powers that be were to hit upon the novel idea of granting a half-share (a 'moiety') of the fine imposed to any person whose information led to a successful conviction. Thereafter none of those who undertook clandestine expeditions to the Funks or any other sea bird breeding ground could be assured that he would not be reported to a law officer by his neighbour. It is regrettable in the extreme that such an expedient was not adopted from the first.

Having corrected the institutionalized abuse of process which had traditionally characterized the administration of justice in the island, it is very much to be doubted if the high-minded Governors of Newfoundland in the late eighteenth and early nineteenth centuries would have stooped to enshrine such a principle in law — even supposing such an idea had occurred to them. Nevertheless, in the absence of a radical reappraisal of how the sea birds should best be exploited to ensure the continuation of a healthy breeding population, nothing less would have sufficed to save the remnant of a once abundant species.

Of course, no one on Newfoundland had any more idea that what was taking place was the effective extermination of an entire species as opposed to a local extinction, than did their contemporaries in Europe. Like Edward Wix, thoughtful people wondered if the Great Auk had not simply sought out a new breeding site, though its absence from the waters between the land and the Banks must have made this seem improbable even at that time. As Sir Richard Bonnycastle phrased it: 'It is of much interest to natural history to know whether it has been really extirpated or has only fled to uninhabited regions from its persecutors, like the Red Indian and the Walrus.'

One of the last recorded glimpses we have of the Beothuk Indians is of a party of them approaching Funk Island in their birch-bark canoes to gather eggs and birds, as they had done since time immemorial, only to encounter a party of Europeans already encamped there. It is no surprise to learn that they were met with a volley of lead shot and were compelled to paddle the thirty-five miles back to Newfoundland empty handed, leaving their assailants to continue their work of destruction without further interruption.

Notes

1 Edward Newton, knighted in 1887, became Colonial Secretary in Mauritius, a post which enabled him to become an authority on the extinct Dodo and to supply his elder brother with the skeletal remains of this bird.

2 The mackerel stocks were, in fact, affected by climatic conditions rather than over-exploitation. The rapid expansion in the numbers of sea birds on the

breeding grounds adjacent to Newfoundland in the mid-twentieth century is probably attributable to the mackerel reverting to their former distribution.

3 This letter is the only evidence that the Great Auk was commercially exploited for its body fat. In the absence of any corroboration perhaps it should not be taken at face value — especially as the writer stated that the birds were released after the fat had been procured. It seems likely that the word, as used in this letter, was originally intended to be a reference to feathers.

4 As all the Auks were still considered as belonging to the genus *Mergus* or, loosely speaking, Divers, this term was probably intended to include Great Auks and Guillemots. A number of the other species mentioned, while common in Britain, have never been known to breed on Funk Island.

5 Louisbourg was taken twice by British forces, once in 1745, after which it was returned, and again in 1758 when a decision was taken to destroy its fortifications.

6 When Peter Stuvitz was in St John's (see Chapter 5) he was puzzled by accounts of the penguins being burned 'for to boil the kettle' and as all the incinerated birds could have been sold as salted meat it does seem a little strange that it should have been thought necessary to dip the birds in scalding water in order to loosen the feathers. Although no evidence has come to light to support a different interpretation to Michael Carroll's, it could be that very hot water was used in order to remove parasites — especially ticks — from the birds which might otherwise continue to infest the feathers when bagged up for shipment.

7 The only inkling any of the Victorian writers received was a report, necessarily thirdhand, from a Newfoundland resident, Joseph Bartlett, who informed Robert Gray, author of *Birds of the West of Scotland* (1871) that he recalled his father, who was born only at the beginning of the nineteenth century, saying that crews used to get on the Funks where they 'built enclosures, lit fires and burnt the birds to death for pure mischief'. Gray passed on this information without comment in a paper entitled 'On the Great Auk' published in the *Proceedings of the Royal Society of Edinburgh* (1879–80). Aaron Thomas's account of Funk Island was first reproduced only in 1899 (see Ussher R. (1899) *Irish Naturalist* **8**, 1–3).

8 The curiously flippant tone of Aaron Thomas' account probably reflects the manner in which the information was given to him: black humour may be interpreted as a defensive psychological reaction to receiving distressing information. The rest of his *Journal* was written in very clear, and rather dignified, English prose.

CHAPTER 11

The old wisdom of the Faeroe Islands

It is at first surprising to find that, in view of the position of the Faeroe Islands as more or less midway between the known breeding grounds of the Great Auk on the Icelandic Geirfuglasker and those on the northern and western islands of Scotland, the record is almost silent in respect of the species breeding there. The remarkable geological formations which may be found on this archipelago, which comprises some twenty-one islands of a total area of 540 square miles, indicate why this was so:

> Most of the islands of which the group is composed are mountains whose foundations are far below the surface of the water, and their sides are divided into horizontal terraces from the bedding of the trap rock. In many places, especially to the north and west, there are precipices of such a stupendous height as to have their summits generally in the clouds, and they are often perpendicular from top to bottom; but they are frequently interrupted by broad grassy ledges upon which the sloping turf is generally undermined by the holes of countless puffins.

These words were written by the youthful English ornithologist John Wolley, in a paper which he read to the natural history section of the British Association for the Advancement of Science following his visit to the Danish dependency in the summer of 1849. Describing[1] the prevailing climatic conditions on the islands, and their topography, he continued:

> Fierce blasts of wind frequently rush down from the mountains, mists and rain are almost incessant, and the air is so damp that the sods of grass with which the roofs of the wooden houses of men are covered are even at the end of summer of the most spring-like green; yet there are seldom any uncomfortably hot days in summer and no very severe cold in winter...Lakes are not numerous but there are a few that satisfy the wants of the common Wild Duck, and also the Red-throated Diver (for trout and salmon are not absent) and they serve as fresh-water baths to continued flocks of kittiwakes. There is no heather that could be sufficient for the Red Grouse; no tree or even shrub of a foot in height.

He then added the following significant words:

> In this brief sketch of the inducements held out to birds to take up their abodes in these islands, we should not omit to notice [i.e. point out] the general peaceable character of the human inhabitants who do not constantly molest them but catch them only at certain seasons and then with as little disturbance as possible.

We need to advance in time a whole century to find a description of the traditional fowling techniques of the Faeroese, which, it is evident, were passed down from father to son with only a minimum of innovation. Another British ornithologist, Kenneth Williamson, who was based on the Faeroes in the Second World War, minutely studied the islanders at their work in catching Puffins from the cliff tops with nets attached to long wooden handles:

> What are the golden rules of this intricate game? First among them, I was told at Mykines is that you must never shift the site of a *fleyging* [hand-netting] place or *rók*. You might imagine that an occasional change would be all to the good, ensuring less exploitation of a particular haunt, and giving the colony there a chance to recuperate. The truth is that a colony has never any need to replenish its strength, because the old wisdom of the Faeroese fowlers sees to it that the vital part of the population — namely the breeding birds — remains untouched. Any bird approaching the land with orderly rows of small fish fringing its coloured bill is allowed to go by unmolested.

An alternative method of taking the birds was to remove them from their burrows early in the season by means of a 'puffin-crook' ('lundakrókur'), leaving sufficient time for the nesting site to be adopted by a younger bird which had not yet acquired a territory. After due time for reflection Williamson remarked:

> It is not likely that this type of fowling, since it takes place on one or two days only at the beginning of the season, harms the breeding potential of large colonies to any extent, although...the method was discontinued on Nólsoy in the middle of the last century because it was thought to be impairing the strength of the colonies there.

It is surprising to learn that these islanders, without driving the Auks to the edge of extinction, were able to supply not only their immediate needs by these methods, but also a thriving bedding industry. Kenneth Williamson related how:

> The soft white breast feathers of the auks are extensively used for stuffing pillows and the thick quilts which form the only bed covers in Faeroe houses, and for many years before the war quantities of these feathers were exported to Denmark for a similar purpose. Plucking is a woman's occupation, and a good worker can deal with three hundred birds a day.

It is probable that the annual cull of Puffins, Guillemots, and Razorbills on the Faeroe Islands, which was conducted systematically and with careful forethought (like that on St Kilda as described by Japetus Steenstrup in Chapter 7) used to be in the order of hundreds of thousands. There is not the slightest suggestion, however, that the long-term ability of any of these species to maintain their numbers was adversely affected.

In his paper read to the British Association a year after his trip to the islands, John Wolley wrote:

> About *Alca impennis* [the Great Auk] I made enquiries wherever I had opportunity but I could learn very little. An old man, Paul Joënsen, had seen one fifty years ago, sitting among the *Hedlafuglur*, that is young guillemots and other birds upon the low rocks, and old men told him it was very rare...Old people have been heard to say that formerly, when many of them were seen, it was considered a sign of a good bird year [a fact which Wolley wrongly, though not unreasonably, explained by supposing] that the same kind of weather which prevented *A. impennis* going to the North, also kept more of his congeners from their far northern breeding places.

As we have seen, the Great Auk was not a bird of the far north, so a different explanation has to be found for the periodic 'Garefowl years' in both the Faeroes and Iceland. As John Wolley implied, albeit unwittingly, the species became very scarce in northern European waters at about the time[2] it was being subjected to the unprecedented destruction instigated by the feather hunters of Newfoundland. This suggests immediately that the European populations of this bird were not isolated, and that their numbers were augmented from time to time by wanderers from the New World — a fact which, if true, speaks volumes for the robustness of this flightless species. It is very likely that these wanderers were birds which had not reached sexual maturity and were biding their time until ready to resort to their ancestral breeding site or sites, most notably Funk Island. If true, this surmise would explain why records of even occasional nesting by the Great Auk in the Faeroe Islands are almost unknown.

Details of the last 'Garfugel' to be captured on the Faeroes were not published until the appearance of H.W. Feilden's *Birds of the Faeroe Islands* in 1872:

> During my recent visit, I landed on the island of Skuoe in company with Herr Sysselmand [i.e. regional administrator] Winther, who took me to the cottage of an old man, Jan Hansen, now 81 years of age. He is believed to be the last man alive in the islands who remembers seeing a garefowl in Faeroe and from a comparison of dates it would appear to be the last recorded instance of the appearance of the bird in these islands; from the attention the natives pay to the arrival and departure of sea-birds, with their frequent visits to the *fuglbergs*, and constant journeyings in boats round the isles, it would be very unlikely for any garefowl since to have escaped observation. This old man who is now blind told me that on the 1st of July 1808 he went with a crew to the Great Dimon[3] for the purpose of catching rock birds: upon a ledge at the base of the cliffs of that island they came across a single *garfuglir*, which was captured; this bird weighed

> 9 Danish pounds [4.5 kg or about 10 English pounds] and on the division of the birds at the conclusion of the fowling, was deemed equivalent to six[4] guillemots.

John Wolley's interest in the Great Auk had been aroused by information he had received from a friend, William Proctor, the keeper of the bird collection at Durham University Museum and a dedicated ornithologist. In 1837 Proctor had returned empty handed from an expedition to the supposed[5] breeding grounds of the Great Auk in Iceland; he seems to have been the first person in Britain to intimate that the species might have become extinct. In 1846, the year he received this information, Wolley had been undertaking extensive research in the British Library into the history of the extinct Dodo[6] and the rumours that Proctor had brought back from Iceland stimulated in him a parallel interest.

Although his research was not as extensive as that of Japetus Steenstrup, whom he was to meet in 1856 in Oslo, Wolley was nevertheless among the first European ornithologists to endeavour to establish the species' range. Accordingly, he noted down from old charts all the islands off Newfoundland and elsewhere which were once Great Auk breeding grounds and entered these on a detailed map of the north Atlantic. He was, however, seriously misled by the erroneous comments which appeared in the most authoritative books published in the early nineteenth century — notably John Gould's *Birds of Europe* (1837) — for he concluded his remarks in his address to the British Association as follows:

Fig. 1 William Proctor (1798–1877) who searched in vain for the Great Auk in Iceland in 1837. From an engraving in the Durham University Museum collection.

> I give all these particulars, as the *Alca impennis* is now looked upon by ornithologists with so much interest as so very rare a bird — so rare indeed, that it has even been suggested that it is extinct. This, however, is not likely to be the case, even without considering the probability of its being found on the Labrador coast.

During the next few years Wolley was to hear and read a sufficient number of reports, albeit uncorroborated, of the continued existence of the species for him to persist in the hope that there were as yet undiscovered breeding grounds where the Great Auk maintained a precarious existence. As his expedition to the Faeroe Islands suggests, Wolley was the most active field ornithologist of his generation. Having decided to give up the study of law in London, he had moved to Edinburgh in 1847 to study medicine. In the Highlands of Scotland, the following year, he is said to have become 'as familiar with the King of Birds [the Golden Eagle] as others are with crows and magpies'. He was able, too, to spend days in succession lying among the breeding sea birds on the Caithness headlands, observing their behaviour through his telescope.

During the next decade he made extensive trips to the subarctic and Arctic regions of Europe in search of the breeding grounds of birds known to contemporary ornithologists, at best, only as passage migrants or winter visitors. These trips were financed by the sale of duplicate sets of eggs and such were the prices paid at auction for those of Gyr Falcon, Crane, Temminck's Stint, Broad-billed Sandpiper, Jack Snipe, Spotted Redshank, Hawk Owl, Shorelark, Bluethroat, Pine Grosbeak, and, above all, the Waxwing (whose nest was discovered by his men for the first time ever in 1856) that the exploits of John Wolley even reached the attention of the European press.

In 1851 Wolley met another great authority on birds' eggs, Alfred Newton, who was some five years younger. Many of the nineteenth-century egg collectors, including Alfred Newton, preferred to style themselves 'oologists' in order to support their contention that the collection and study of birds' eggs was a genuine branch of science and not merely a collecting mania. Newton, for instance, on the basis of a single unmarked egg taken on the Farne Islands, had argued in *The Zoologist* magazine what was widely believed at the time, namely that the bridled (or 'ringed') morph of the Common Guillemot was an entirely separate species. By way of a reply John Wolley brought all his considerable field experience to bear in exposing what he evidently considered to be unscientific nonsense:

> In the Faeroe Islands the ringed guillemots struck me as being perhaps in greater plenty than in the north of Scotland especially on the little rocks at the level of the sea; of course not breeding in those low situations; but the common guillemots were always in far greater numbers than the ringed, and always mixed with them. Down the stupendous cliffs of these islands, I did not attend so much to the guillemots when such rare and interesting birds as the fulmars fully occupied me, but I examined the heaps of broken-necked birds brought up by the climbers for

> provision; here, if I remember right, the proportion was about one to ten. Of two ringed birds which I dissected in Fugloe, one was a male, the other a female. I constantly made enquiries of the people, who are very intelligent and very intimately acquainted with their birds. They none of them have ever dreamed of the white-eyed birds being a different kind from the others, but some of them thought they were the males, others that they were the females — both opinions, as I ascertained, only partially true...I feel convinced, that if the ornithologists who have described the two species of guillemot had had opportunities of seeing them on their native rocks, the idea of their being distinct could hardly have occurred to them.

The last sentence seems a mild enough rebuke of the 'armchair' naturalists, and it was not intended to include among that class Alfred Newton himself. Despite having to use walking sticks all through his life on account of an injury sustained in childhood, Newton was as active as anyone in that condition possibly could be.[7] Nevertheless, he would have been only human if he had found Wolley's last remark a painful reminder that he was not in a fit condition to make some of the firsthand observations that his older contemporary was able to. However, recognizing each other as among the leading ornithologists of their day, Wolley and Newton struck up quite a friendship. Being young, both had limitless confidence in their abilities and in the future, which would, surely, shower them with honours.

During his stay on the Faeroes in 1849, John Wolley formed an abiding impression of the relationship which existed between men and birds on the islands. The final paragraph of the paper which he read to the British Association in 1850 was something of a clarion call:

> We cannot conclude this subject better than by deriving a lesson from the simple and happy people of the Faeroe Islands in their treatment of the birds which surround them. It is a melancholy thing to see how at almost all the great breeding places of sea-fowl round Britain the numbers are rapidly diminishing every year in consequence of the pitiless persecution which afflicts them — slaughtering parties visit them by trainfuls. The rights of the bird climbers established by long usage, require the assistance of the law; and all persons concerned in the coast navigation, should interest themselves to procure, by Act of Parliament or otherwise, protection for sea-fowl at those places, where at that time of the year when they throw themselves entirely upon the mercy of mankind, for by such protection alone can their sure and speedy extirpation be averted; and one of the best kind of beacons, the flight and clamour of birds, be preserved to warn vessels in foggy weather of their approach to the dangerous headlands of our coast...The numbers are not seriously lessened by legitimate bird-catchers any more than in the case of poultry in a farmyard.

Although it was to take nearly twenty years to achieve, and John Wolley was not to live to see it, the statutory protection of sea birds was eventually accom-

Fig. 2 John Wolley.

plished. The fact that John Wolley was only twenty-five when he wrote the paper from which the last paragraph has been taken, indicates, if any further evidence were needed, not only the boldness and originality of his mind, but also the degree of authority he seemed to exercise in the matters on which he wrote — an authority gained from his unrivalled experience as a field ornithologist.

Notes

1 The Linnaean names given by Wolley have been omitted as these have, in most cases, been superseded.

2 The information given by Wolley is fully corroborated by Landt, writing *c.*1800, who stated, '[T]he garfuglen begin to become rare.' This statement may be contrasted with that of another naturalist, Mohr, who stated simply in his *Natural History of Iceland* (1780) that a number of Garefowl were found among the other sea fowl on the Faeroes. According to another writer, Jens-Christian Swabo, this same Mohr received the egg of a Great Auk

removed from the oviduct of a bird killed on Fugloe, in the north of the archipelago (unpublished MS, Royal Library, Copenhagen). There is, in addition, a hearsay account given by C.-J. Graba in his *Diary Kept During a Trip to the Faeroes in 1828* (1830), supposedly related by an old man on one of the islands, to the effect that an incubating bird had been taken at Westmannahavn in the Faeroes; however, as the details sound strikingly similar to the account given by the principal fowler on the Westmann Islands off southern Iceland to Friedrich Faber and published by him in 1822, it must remain a dubious record. Graba seems to have repeated Benicken's assertion made in 1824 that the Great Auk should be considered extinct in northern European waters.

3 Kenneth Williamson gave the name of the site as 'the rocks of Stóradífun'.

4 John Wolley, writing a generation earlier, gave the traditional equivalent as 'four *lombvias* or Guillemots.'

5 In fact Proctor had been searching on the island of Grimsey, just above the Arctic Circle, which was outside the normal range of the species.

6 John Wolley made over the information he had discovered to H.E. Strickland who incorporated it in *The Dodo and Its Kindred* (1848).

7. In 1855 Alfred Newton became the first European ornithologist to discover, in Norwegian Lapland, the nest of the Red-throated Pipit.

CHAPTER 12

In search of the Great Auk

When, in the midsummer of 1855, Alfred Newton[1] joined John Wolley at Vadsø, at the head of the Varanger Fjord, East Finmark, he found his friend in a state of some excitement having heard a convincing firsthand account, from a local man, of the shooting of a Great Auk (one of four observed off nearby Vardø in April 1848). The body of the bird, all wet and bloody, had been deposited on the shore, but by the following morning it had been washed out to sea in an unexpected storm during the night (see Chapter 13). Despite the passage of time Wolley, quite characteristically, had had no hesitation in instituting a search just in case the skeletal remains of this bird, or another, could be found.

John Wolley was not alone in refusing to join the increasing body of opinion which believed the Great Auk to be extinct. In 1854 an excerpt from the *Scandinavian Adventures* of Lewis Lloyd appeared in *The Edinburgh New Philosophical Journal* under the sensational headline 'The Great Auk Still Found In Iceland'. Basing his information on the *Ornithologia Danica* of Niels Kjærbölling (1851), Lloyd gave an incomplete list of Great Auks killed in Icelandic waters from 1813 to 1844 with the added rider 'people whose word is to be relied on, Kjærbolling tells us, have informed him that birds have subsequently been seen off the coast of Iceland'. The article concluded: 'From the above account there can be little question as to the Great Auk still existing in some numbers on the coast of Iceland; and I doubt not that we shall one day hear of some of our enterprising countrymen having overcome all difficulties, and returning home with a rich booty.'

Taking up the challenge, Wolley and Newton made the decision while at Vadsø to visit the Icelandic bird-skerries, as soon as feasible, to ascertain whether the Great Auk could yet be found there — and, there can be no doubt, to enhance if possible their already prestigious egg collections. A year before the extract from Lewis Lloyd's book was published, a much gloomier view had been taken by another writer, James Wilson, in the *North British Review* (May 1853). 'Many considerate people begin to question the continuance of its existence upon earth,' he wrote and went on to cite the 'absence of records from its breeding stations in recent years'. Even if others, including Wolley and Newton, had been justified in believing in the species' continued survival on imagined breeding grounds in Labrador and Spitzbergen, there was already compelling evidence that in Europe, at least, the Great Auk was, if not extinct, then at best on the verge of extinction.

In the light of this it is not difficult to see that any pressure from collectors would have been intolerable for such a species as the Great Auk. By the time that Wolley and Newton made up their minds to go to Iceland, there can have been no excuse on the grounds of ignorance. Whatever the separate roles that these two friends would in due course play in bringing about the statutory protection of wild birds, ironically, in respect of the Great Auk, we have to conclude that both were part of the problem — not its solution.

From 21 May to 14 July 1858, with the exception of two short intervals for excursions, the two friends boarded at the home of Vilhjálmur Hákonarsson, near Kyrkjuvogr, a sheltered bay on the Reykjanes promontory lying within striking distance of both Eldey and those more distant bird-skerries which had been visited by Friedrich Faber ('Fugl Faber' as the Icelanders had nicknamed

Fig. 1 Alfred Newton.

him) a generation previously. Their host, known as 'Geirzoega' (i.e. 'Gare-leader'), who fortunately — and for that time perhaps, remarkably — had a good knowledge of English, had led the last successful expedition after Great Auks and was anxious to make another visit. The weather, however, was against them and, in Alfred Newton's words: 'boats and men were engaged, and stores for the trip laid in; but not a single opportunity occurred when a landing would have been practicable. I may say it was with heavy hearts we witnessed the season wearing away without giving us the wished for chance.'

Although Eldey is precipitous and all but inaccessible to man, intrepid fowlers had succeeded in driving large pegs, or nails, into the rockface to serve as footholds and it was here that a certain Brandur Guðmundsson, from the same village, discovered thirty or forty breeding Great Auks in 1830. These birds, which traditionally had frequented the Geirfuglasker some fourteen miles further out, had adopted Eldey when their ancestral breeding ground was submerged in the volcanic disturbance in March of that year. Usually regarded as an unsuitable breeding site for Great Auks, Eldey does in fact have one point of access where the birds could come ashore — a slope known as the Underland, reaching up to the base of the towering cliffs of the island. Whether or not the Icelandic breeding population could have survived if the Geirfuglasker had not disappeared remains an open question; the likelihood is however that the financial inducements[2] held out to Icelanders to overcome their fear of venturing out to the more distant bird-skerries would have been increased sufficiently for them to risk 'putting life and death on an even chance' to obtain the birds. While this is a matter of conjecture, and whilst we should not forget that no one realized how few Great Auks there were left, it is difficult to disagree with the view expressed by the naturalist and writer Richard Perry that it was 'typical of the age that all the Eldey survivors had been collected for museums within fourteen years of the eruption'.

Faced with the prospect of enforced idleness, John Wolley put his energy and evident knowledge of the Norse languages to good effect, buttonholing all the survivors of earlier 'Geirfugl' expeditions and anyone else who had firsthand knowledge of the bird. He ascertained little about the old Geirfuglasker beyond the fact that it had sunk on or about 6 March 1830, that it was visible on clear days from Stafness on the peninsula, and that its shape was 'just like Zoega's wide-awake hat, high in the middle with low land all round'. Wolley's informant, Erlendur Guðmundsson, aged eighty-one and described as 'a remarkable man', stated that it must have been forty years since anyone had gone out to the Geirfugl-skerry. Another elderly witness, Eiríkur Gunnlaugsson, said that when he was a boy they 'never, never, never' went out there on account of a superstitious dread of the place. The story was that a certain Thorwaldur, in the fourteenth century, had been left on the island and had survived the winter but had refused to tell those who found him in the spring how he had done

Fig. 2 Eldey or 'The Mealsack'. (Photograph by Gunnar Hallgrímsson)

so. The myth developed that he had been looked after by the 'Hyldafolk' ('fairies and ghosts' said Vilhjálmur). Later in life Thorwaldur is said to have gone insane, deserted his farm at Mellaberg, and to have thrown himself off a cliff into the sea (where, some said, he had turned into a whale).

The impact on others of such stories may perhaps be gauged from one particular episode, narrated by Erlendur Guðmundsson, relating to the same legend. Early in the summer after Thorwaldur's return a cradle with a child in it was left inside the church door at Hvalness:

> The priest asked if any man in the church would ask for the child to be baptised; no-one would — the priest, finding no-one, did not baptise the child. There [was] layd over it a red coverlet — after a little time he took hold of it & drew it away & the child was gone. Of this coverlet was made a covering for the altar & also a dress for the priest...It was supposed that the child came from the Geirfuglasker, from some being there. (Hyldafolk.)

No doubt stories of this kind helped to ensure that the number of people who made regular trips to the skerry were few and thus, they account in part for the survival of the small colony of Great Auks on the Geirfuglasker for a generation after the extirpation of the principal breeding population in the New World. The discovery in 1732 of the skeletal remains of a fowling party who had been stranded on the island would have done nothing to diminish its reputation as a place to be feared.

Following the eruption which sank the Geirfuglasker, more than forty specimens were taken from Eldey between 1830 and 1831, and a dozen or more in 1833. It is not known how many other expeditions took place over the next decade — certainly in 1836 a French ship landed a crew who made off with five Great Auks — but it is clear that the number of birds taken in any one year during that period never reached double figures. Behind the depredations carried out by the villagers of the Reykjanes promontory we can detect the instigating hand of the merchants of Keflavik and Reykjavik and even from Denmark who supplied dealers in Hamburg or sent specimens direct to the Royal Danish Museum at Copenhagen. One or two naturalists in Iceland obtained Great Auks at this time; one such, Herr Mechlenburg, forwarded specimens to collectors in other parts of Europe including one, together with an egg it was said to have laid, to the British naturalist John Hancock who established the Newcastle Museum.

The capture of what are regarded as the last two specimens of the Great Auk came about in the following way. Vilhjálmur Hákonarsson, to whom the idea of another expedition had been suggested by a dealer, Carl F. Siemsen of Reykjavik, organized a party comprising fourteen men, including himself, to row out to Eldey at the appropriate time. They departed in an eight-oared boat (which, it will be recalled, was necessary in case they encountered adverse weather conditions) one evening during or soon after the 'Fardagar' from the 3rd to the 5th of June in, according to Alfred Newton, 1844. As the crew neared the island next morning they could plainly see two Great Auks on the highest part of the Underland. To be quite sure of securing them, Hákonarsson intended to put four men ashore, but as the weather had begun to deteriorate, one of those who had been designated refused point-blank to take any further part. The three men who got up on to the island raced up towards the two birds on the slope. Jón Brandsson, a veteran of previous Garefowl expeditions, secured one of them while the other was pursued by Sigurður Iselfsson and Ketil Ketilsson to the edge of a precipice where, Ketil Ketilsson's nerve failing him as he neared the cliff top, it was left to Sigurður Iselfsson to seize the bird just before it leapt down to the sea. Ketilsson walked back 'between 20 and 30 fathoms' to where the birds had started from when the men had got on to the Underland. There he found a single egg on the flat lava rock. He claimed later that on picking it up he saw it had been cracked and so put it down again. It is more likely that he dropped it but was too embarrassed to say so.

John Wolley and Alfred Newton received some conflicting information about the number of eggs found on Eldey during this last Garefowl expedition. In a letter to his friend Thomas Southwell, written on August 30th after his return home, Newton stated: 'If, as one man says, there were two, there must have been other birds out fishing at the time the boat landed and then there is a strong presumption that the bird was not exterminated in that year.' It is clear from John Wolley's *Garefowl Book* preserved at Cambridge University that the 'one man' referred to by Newton was none other than the skipper of those expeditions, who had been Newton and Wolley's chief helper in their enquiries, as well as their host — Vilhjálmur Hákonarsson.

Whether, when these two birds were strangled, they represented the last of their species is debatable. Subsequently, Alfred Newton was to declare his personal belief or, at least, hope that 'there may be still "some happier island in the watery waste" to which the Penguins of the western seas may have escaped' adding 'but then, we may rely upon it, there is left a scanty remnant only'.[3] We may wonder how the belief that the species finally became extinct at a definable point in time — a belief which has ever been uncritically repeated by later commentators[4] — has existed so long without being challenged. After all, Kjærbölling had stated that people whose word he trusted in matters of this kind had assured him that there had indeed been sightings after 1844 — a contention which finds support in an often overlooked piece of information sent *in litt.* to the Revd J.G. Wood, the editor of *The Illustrated Natural History* (1862): 'Mr David Graham [of the York Museum] mentions that in 1846 he met a fisherman in Iceland who had two Great Auks and two eggs which he offered for 20*l.*, an offer which was refused to the writer's subsequent regret.'

All that can be said with certainty of the two birds killed on Eldey by Jón Brandsson and Sigurður Iselfsson is that they represent the last incontrovertible record of the species which, represented by a handful of birds, might well have continued for some years more. If this seems a fanciful suggestion, it only needs to be asked whether or not it is altogether improbable that a small number of non-breeding birds were still in existence at the time that the last authenticated pair[5] was taken.

In fact, it is not certain that the last expedition to capture geirfuglar took place in June 1844. Much of the verbal testimony so carefully written down by John Wolley in 1858 is, as Alfred Newton subsequently observed, conflicting and irreconcilable. This should not be surprising for, notwithstanding their deserved reputation for truthfulness, Wolley's witnesses — men and women who wrought their living from soil and sea — are known on occasion to have had to check their parish records to be quite sure how old they were. Ketil Ketilsson told John Wolley he believed the expedition in which he partook occurred in 1846, whilst Vilhjálmur Hákonarsson, the *formaður* (skipper), on the grounds that there were five years between the two last expeditions to go in search of Great Auks and that that last, unsuccessful, expedition had been only four years prior to the arrival of John Wolley and Alfred Newton in Iceland, was adamant that 1849 was the true date.

The only objective evidence comes from Iselfsson's employers who calculated that the last year in which their man could have taken part in such an expedition was 1844 — a year or two before they engaged his services. It seems that Alfred Newton, when working on an abstract of John Wolley's researches for publication in the journal *The Ibis*, regarded this testimony as decisive. It is more than probable, however, that there were *two* occasions on which two birds were taken and that, in the intervening years, the details of these became confused in the minds of those who had knowledge of both. This finds corroboration in the fact that Hákonarsson gave the names of two of the three men to be landed on Eldey as Frímann Gislason and Martin Ólafsson. Jón Brandsson, moreover, told John Wolley that while he had taken one bird he could not recall whether it was 'Ketil's son Ketil or Frímann Gislason' who had taken the other.

If, as there seems no real reason to doubt, an expedition which secured two birds did indeed set out in June 1844, it seems likely that there was another subsequently with a similar outcome. If so, it might explain the conflicting testimony over the number of eggs found on the last occasion on which birds were secured.

William Proctor's expedition to Iceland in 1837, which was unsuccessful as regards finding the Great Auk (see Chapter 11), was cited in W.C. Hewitson's *Coloured Illustrations of the Eggs of British Birds* (1856) as evidence of the species' probable extinction. When, in the 1860s, it was finally accepted by all but a tiny minority of commentators that Iceland's Geirfugl was no more, the belief held by some that the species had died out in or about 1844 seemed to

Fig. 3 The last of its kind or were there non-breeding birds out at sea?

have been wholly justified. This belief, moreover, appears to have acquired the status of a cherished notion among a good many readers of popular accounts of the fate of the species, and so became progressively more difficult for even the most iconoclastic of writers to call into question.

It is ironic that some of the post-1844 records of sightings (to be discussed in the following chapter) are rather more credible than some of those pre-dating 1844 which *are* accepted. In the case of the latter, acceptance is presumably based on the dubious grounds that less rigour need be applied in deciding whether to accept such a record when there is no question of the species' having entirely disappeared. With the benefit of greater hindsight it is wiser to date the species' ultimate extinction no more precisely than to the middle decade of the nineteenth century and, in so doing, remove from Messrs. Brandsson, Iselfsson, and Ketilsson the opprobrium of having been the unwitting exterminators of a species — for the misdeeds of these three seem quite venial compared to those of the anonymous Newfoundland men whose reeking pyres on Funk Island had sent heavenwards an unholy burnt offering.

During their stay, Alfred Newton and John Wolley intended to reach a number of possible Great Auk breeding sites — though not, apparently, the Geirfuglasker off the Westmann Islands. Among these they listed a 'Geirfugladrángur', which was in fact none other than the steep-sided Grenadier's Cap[6] described by Friedrich Faber, on which they erroneously thought Count Raben had landed in 1821. In addition, on Olsson's map which Wolley had consulted prior to leaving Britain, was another Geirfuglasker, named Hvalsbak (pronounced as 'Hqualsbak' and meaning 'the Whale's Back') by Danish fishermen, lying some forty miles off Breiðdalsvík on the south east coast of Iceland. In the belief that each island on which the Great Auk bred held its own discrete colony (rather than there being a local Icelandic population which tended to colonize smaller islands only when the breeding population was relatively high), John Wolley and Alfred Newton went to considerable expense in engaging the services of a young theology student, Eiríkur Magnusson, to charter a boat to go out to the island on their behalf, leaving them free to concentrate their efforts in the south west of the country. Alfred Newton believed this island to be 'the best card in the pack and one I should never forgive myself for not playing if afterwards it should turn out the birds were there'. Not surprisingly, however, their young friend returned with 'negative news', commenting that there was no local tradition regarding the breeding of the Geirfugl on that island, at least not within living memory.

It is remarkable how little the Icelanders of the fishing communities on the Reykjanes peninsula knew about the natural history of the Geirfugl: 'All accounts agree,' wrote Alfred Newton, 'in saying that on land the bird is blind and only gets its sight when it is in the water, but it has capital ears' — which seemed to be a way of saying that 'they were easily frightened by noise, but not

Fig. 4 Aerial view of Eldey and Cape Reykjanes. The meeting of cold northern and warmer southern waters, creating a dangerous roost, can be seen clearly. (Photograph by the author)

by what they saw'. This widespread misapprehension seems to have been caused by the evident conspicuousness on the Great Auk of the nictitating membrane ('blaðka'), which serves to protect a bird's eye. The other details that were ascertained and published in *The Ibis* (1861) amount to no more than a thumbnail sketch:

> We were told by many people that they swam with their heads much lifted up, but their necks drawn in; they never tried to flap along the water, but dived as soon as alarmed. On the rocks they sat more upright than either guillemots or razorbills, and their station was further removed from the sea...They sometimes uttered a few low croaks.[7] They have never been known to defend their eggs, but would bite fiercely if they had the chance when caught. They walk or run with little, short steps, and go straight like a man. One has been known to drop down some two fathoms off the rock into the water. Finally, I may add that the colour of the inside of their mouths is said to have been yellow as in the allied species.

There were, Newton stated, 'many other particulars of interest' which, but for lack of space, he could have given. The testimony of two of John Wolley's witnesses, Guðni Hákonarsson and Jón Gunnarsson, deserve mention as they relate to the very damaging expeditions in the years immediately after the disappearance of the Geirfuglasker nearly thirty years previously. The frequent use of the present tense reveals the confidence many Icelanders had that the Great Auk still bred on Eldey and that, weather permitting, Newton and Wolley had a good chance of finding them there. Guðni Hákonarsson recalled that the Great Auk:

> ...sits up just like svartfugle [Guillemots] close to the cliff always on the same flat stone shelf close to the back-cliff — breast always to the sea. When one comes up the rock the other birds begin to cry, and the Geirfowl comes down to meet one — comes walking slowly (Guðni makes action) like children but quite upright — neck straight up. Guðni speaks this last with emphasis, decidedly, as though he remembers well.

Jón Gunnarsson told Wolley how:

> The Geirfugl is frightened when he hears the Gannet croaking. He then begins to hear round him — (Pantomime shewn of men going with short silent steps to catch the bird, stooping, arms bowed)...As one approached the bird it began to walk a little about its egg and to turn its head about — It was quite silent — but then it was seized round the neck.

Evidently by the 1840s the birds had become much more wary, leaving their eggs as soon as anyone set foot on the rock.

In his letter to Thomas Southwell written from home at the end of August, Newton addressed himself to the question of how it was that the Great Auk seemed to have ceased to exist:

> As to the extinction of the Great Auk, if it is extinct I think it has been mainly accomplished by human means...Under the influence of the 'Almighty Dollar' (tho' in Iceland it is not worth 4s. 2d.) these poor birds were persecuted, their eggs plundered and their necks broken[8] to supply the demand which museums were then creating. And so the number dwindled, until in 1844, the only two then to be seen were taken, their egg broken (the shell left on the rock) and their skins shipped to Europe. I do not think there is any good evidence of the bird being seen since that time; but I confess I do not give it quite up, nor shall I for the next five or six years, though the places suitable for its breeding station must be very few in number.

The view expressed about the role of the 'Almighty Dollar' in the demand-led exploitation, or perhaps we should say plunder, of the last remaining Great Auk colony on Eldey would seem to be little more than high Tory disdain for 'Yankee commercialism'.[9] The truth is that American museums only sought to purchase examples of the Great Auk once it was realized that the species was extinct and had become so under the noses, almost, of American naturalists.[10] By contrast, the Royal Danish Museum at Copenhagen continued, for many years after the species' presumed extinction, to offer a large reward to anyone in the Danish dominions who secured a specimen.

Unsubstantiated, and presumably apocryphal, reports of the occasional Great Auk in Greenland's territorial waters were received even as late as the late 1860s and Mr Frederik Hansen, the Governor of Godhavn on Disko island, who was described as an enthusiastic naturalist, was said to have been 'very sanguine' at the time that he would obtain one.[11] Likewise, Sir Leopold McLintock, renowned for his part in the search for the ill-fated Franklin expedition in Arctic Canada, had made diligent enquiries as to the occurrence of the Great Auk in Greenland: 'The resident Europeans,' he wrote, 'are quite aware of the value of this bird and keep a sharp look out for it'. The truth is that there continued to be sufficient demand for specimens of the Great Auk in Europe, even when it was known to be either extinct or very nearly so, for the price to remain high. Consequently it is not necessary to put into the balance the minimal demand which may have come from American collectors or institutions at the time that the species actually became extinct.

Soon after both men had returned to England, John Wolley alarmed his friends by displaying a wholly uncharacteristic lassitude and an increasing loss of memory. On consulting a specialist in the following New Year, he was diagnosed as suffering from an 'affection of the brain' which was incurable and which would, in time, prove fatal. John Wolley, the most energetic British ornithologist of his own and quite possibly of any other generation, died in November 1859, at the age of only thirty-five. Alfred Newton, to whom Wolley had left his enormous collection of birds' eggs, wrote a memoir of him in *The Ibis* (1860) which he concluded with the following tribute:

> His good qualities are treasured in the recollection of those who knew him and especially of one to whom he gave the last token of his esteem, and who, having endeavoured (how imperfectly no-one knows better than himself) to discharge a duty to a deeply lamented comrade, cannot conclude this sketch without a expression of gratitude at having been permitted to share so largely the intimacy of such an upright man.

It may not be entirely fanciful to regard the work carried on in the cause of bird protection as Alfred Newton's lasting memorial to his late friend who, as we have seen, had been its first advocate.

Notes

1 Newton was accompanied by another ornithologist, W.H. Simpson.

2 The value of the birds increased from two or three rix-dollars per bird in the early 1830s to forty per bird twelve years later.

3 Subsequently, Newton postulated the Virgin Rocks near Newfoundland's fishing banks as a site where the Great Auk might have continued to breed in the 1850s.

4 Among such commentators we must include James Fisher writing in Volume I of *Bird Recognition* (1947) and, more recently, Professor Tim Birkhead writing in the *New Scientist*, 28 May 1994. The latter purports to give the exact day of the month on which the Great Auk became extinct.

5 The bodies of two birds were transmitted by Etatsraad (Councillor) Eschricht from his Zootomical Museum to the Zoological Museum attached to Copenhagen University, where, preserved in alcohol, they may be seen to this day. Professor Jon Fjeldså of the Zoological Museum has informed the present writer *in litt.* that the label on the jar, which reads 'Geirfuglaskér [*sic*] Iceland 1844', may have been written later as there appears to be no entry in the museum's accession catalogue giving the date when the specimens were received. Japetus Steenstrup's assertion, based on seeing the specimens, that the birds constituted a pair and were not two females is the only statement which may be taken as evidence, rather than supposition, in the entire literature of the Great Auk. Dissection of the birds was not commenced until 1940 when it was confirmed that they are male and female.

6 Alfred Newton persisted in his belief, first stated in his article in *The Ibis* (1861), that this spot, which was, evidently, the goal of his expedition with John Wolley, might have held a tiny population of *Geirfuglar*. He repeated his assertion in a section on the ornithology of Iceland in S. Baring-Gould's *Iceland: its Scenes and Sagas* (1863) and in an article on the natural history of the same country in *The Zoologist* in the same year. In the last he wrote: '[T]he expedition is one of no small danger and during the last five years has not admitted of its being undertaken. I can only express my sincere wish that whenever this rock is reached, the bold adventurer may reap a fitting reward; but at the same time I must declare my hope that in such a case he will be careful to see that the best possible use is made of the spoils. The mere addition to the already considerable number of stuffed skins and blown eggshells of the species which are dispersed in various collections will be no addition whatever to science. If the bird is doomed to extinction, and such, I fear, is its fate, all who are concerned in bringing about the catastrophe are bound to see that the most is made of whatever Chance may throw in their way.' It is clear from his article in *The Ibis* that what he had in mind was nothing less than a Great Auk vivarium in the Regent's Park Zoo — an idea derived from John Wolley's plan for an eagle aviary there, to which the latter had actually contributed a pair of White-tailed Eagles which he had captured during his tour of Northern Scotland, the Orkneys, and the Shetland Islands in 1848.

7 '...uttered a few low croaks' is Alfred Newton's translation of the expressive Icelandic 'mikið láu kraki.

8 Those who gave evidence to John Wolley without exception stated that the birds were killed by being strangled, one witness actually emphasizing that their necks were not broken.

9 On hearing that a licence had been granted in 1862 to an American entrepreneur by the name of Glyndon to extract 'guano' for use as agricultural fertilizer from Funk Island, Alfred Newton spoke disapprovingly, in an address on 'The Zoology of Ancient Europe', of the advent of a 'Yankee speculator' to the resting place of countless Great Auks. As a comparative anatomist he was particularly concerned as there were no complete skeletons of the species in any of the world's collections at that time. However, Alfred Newton's tone changed when he became the beneficiary of this activity, as he explained the following year: 'The Bishop [of Newfoundland], through Mr N.R. Vail, a gentleman of the United States well-informed on scientific subjects, and therefore aware of the interesting nature of the research, made application to the lessee of Funk Island, who ordered his men employed there to use their best endeavours to obtain for me the bones of the Penguin. They appear to have done their work very effectually, for I hear that they brought away many puncheons of bones and other remains' ('On a Natural Mummy of the Great Auk', *Proceedings of the Zoological Society*, 1863).

10 A specimen of the Great Auk, in the Smithsonian Institution (Catalogue No. 57338), which was taken on Eldey in June 1834, was purchased much later in the century from Herr Wilhelm Schluter of Halle in Germany. It is of interest to observe the number of hands through which the bird passed. The immediate purchaser in Iceland was a merchant named Dethlef Thomsen whose wife and sister-in-law skinned and stuffed the bird and the eight others killed at the time. They were sold then to Herr Liagre, a dealer in Hamburg, who disposed of the bird now in the Smithsonian Institution to another Hamburg dealer, Herr Salmin. In due course, the bird was sold to Herr Goetz in Dresden from whom Herr Schluter acquired it. We may suppose that the price increased with each transaction.

11 R. Brown: 'The Mammals of Greenland', *Proceedings of the Zoological Society*, 1868.

CHAPTER 13

Last appearances

In May 1834, a decade or so before Vilhjálmur Hákonarsson made his last successful excursion to Eldey, an Irish fisherman succeeded in catching a Great Auk, described subsequently as half-starved, by means of a handful of sprats as it neared his boat in Waterford harbour. This bird, the last unquestionable record of the species in the waters of the British Isles, after being kept[1] as a pet for some months, was eventually preserved and mounted by Dr Burkitt who generously donated it to Trinity College, Dublin. It is far from certain, however, that it is the *very* last British or Irish record, for two subsequent sightings of what may well have been the species, are recorded — one generally accepted, the other not.

A good many of the later records of the Great Auk are based on hearsay only, though this in itself should not automatically preclude them from serious consideration. For instance, the simple sentence, 'One was seen off Fair Isle in June, 1798' in the *Historia Naturalis Orcadensis* (1848) by W.B. Baikie and R. Heddle, coming as it does from a part of the world where the Great Northern Diver (the only species with which the Great Auk was likely to be confused) is a well-known visitor, has never been called into question in spite of the absence of corroborative information. From the opposite end of the country (where, again, the Great Northern Diver is a regular winter visitor) comes the following record from Dr Edward Moore's *Catalogue of the Web-footed Birds of Devonshire*: '*Alca impennis.* Great Auk or Penguin — Mr Gosling of Leigham, informed me that a specimen[2] of this bird was picked up dead near Lundy, in the year 1829.'

Authentic or not, this Lundy record had a remarkable legacy. In 1865, William D'Urban and the Revd Murray Mathew, who were in the course of preparing their authoritative *Birds of Devon* (1895), wrote to the Revd H.G. Heaven, a thoroughly competent amateur ornithologist, to request any information he might have regarding this species. They received the following startling reply dated September 6th of that year:

> With regard to your question whether we have ever *seen* the Great Auk, I must answer in the negative. There is a strong presumptive evidence, however, that the

> Great Auk has been seen *alive* on the island in the last thirty years; at least, I cannot imagine what other bird it was. The facts are as follows, and I must leave it to more experienced ornithologists to draw the conclusion: in the year 1838 or '39 as nearly as I can recollect, not, however, more recently, one of our men in the egging season brought us an enormous egg, which we took for an abnormal specimen of the Guillemot's egg, or, as they are locally termed, the 'Picked-billed Murr'. This, however, the man strenuously denied, saying it was the egg of the 'King and Queen Murr' and that it was very rare to get them, as there were only two or three 'King and Queen Murrs' ever on the island. On being further questioned he said they were not like the 'Picked-bills', but like the 'Razor-billed Murrs' (i.e. the Razor-billed Auk); that they were very much larger than either of them; that he did not think they could fly, as he never saw them on the wing nor high up the cliffs like the other birds, and that they, as he expressed it, 'scuttled' into the water, tumbling among the boulders, the egg being only a little way above high water. He thought they had deserted the island, as he had not seen them or an egg for (I believe) fifteen years till the one he brought us; but that they (i.e. the people of the island) sometimes saw nothing of them for four or five years, but he accounted for this by supposing that the birds had fixed on a spot inaccessible to the eggers from the land for breeding purposes.
>
> The shell of the egg we kept for some years, but unfortunately it at last got broken. It was precisely like the Guillemot's egg in shape, nearly, if not quite, twice the size, with white ground and black and brown spots and blotches. We have never, however, met with bird or egg since, but as the island has become, since that time, constantly and yearly more frequented and populous, it may have permanently deserted the place.
>
> The man has been dead for some years now, being then past middle age, and I think he had been an inhabitant of the island some twenty five or thirty years. He spoke of the birds in such a way that one felt convinced of their existence, and that he himself had seen them; but he evidently knew no other name for them than 'King and Queen Murrs', which he said the islanders called them 'because they were so big and stood up so bold like'. In colour they were also like the 'Razor-billed Murr'. Nobody, he said, had ever succeeded in catching or destroying a bird, as far as he knew, because they were so close to the water and scuttled in so fast. The existence of these birds had been traditional on the island when he came to it, and even the oldest inhabitants agreed that there were never more than two or three couples. He himself never knew of more than one couple at a time.

Either this account is the unalloyed truth or Mr Heaven, when a lad of tender years, was the victim of a pointless hoax perpetrated not by one man only, but by an entire island community. D'Urban and Mathew, ignoring the point just made, took shelter behind the notion of a double-yolked Guillemot's egg — a phenomenon which has, very occasionally, been recorded. However, the facts of the account are consistent with what we now know of the decline of the Great Auk on the eastern side of the Atlantic: during the twenty-five years prior to

Fig. 1 Lundy: a huge block of granite rising up from the sea. (Photograph by Alan Richardson)

Fig. 2 Part of the west coast of Lundy. (Photograph by Alan Richardson)

1838–9, occasional pairs bred, apparently at long intervals, while previous to that time (that is until about 1813) up to three pairs were known to breed (again, somewhat irregularly). Quite conceivably, then, the Great Auk was a British breeding bird until a few short years before its extinction.

In July 1840, John Macgillivray, son of J.J. Audubon's friend, Professor William Macgillivray, and a first-rate ornithologist in his own right, made a birding tour of the Outer Hebrides. In his report to William Yarrell (author of the *History of British Birds*, the standard reference work of that era), Macgillivray stated : 'The Great Auk was declared by several of the inhabitants to be not of unfrequent occurrence about St Kilda, where, however, it has not been known to breed for many years back. Three or four specimens only have ever been procured during the memory of the oldest inhabitant.'

Evidently John Macgillivray did not hear an account of the capture and destruction of a bird, sounding suspiciously like a Great Auk, at about the time he visited the Western Isles. Symington Grieve, in his monograph on the species (1885), reproduced a letter from a frequent visitor to St Kilda, Henry Evans, who described what happened:

> A Laughlan McKinnon, now aged 75, tells me that with his father-in-law and Donald McQueen,[3] he caught and killed a garefowl on Stac-an-Armin. He dates the event at about forty years ago. Donald McQueen's son also says it was about forty years ago or perhaps a year or two more. Laughlan McKinnon is the only survivor of the three men. I know him quite well: he appears as bright and sharp as any man in St Kilda. He recognised at once as the Garefowl a picture of the Great Auk I showed him. He especially called attention to the little wings for so large a bird, and to the white spot on the side of the head which he remembers was on the bird; he spoke much of the great bill the bird had which he said it kept open very long and often, 'as if it would never shut its bill again.' He also put both hands to his sides, and said the bird was very flat and fleshy there.

J.A. Harvie-Brown elicited further details which he published in his *Vertebrate Fauna of the Outer Hebrides* (1888):

> McKinnon told Mr Evans that they found the bird on a ledge of rock, that they caught it asleep, tied its legs together, took it up to their bothy, kept it alive for three days, and then killed it with a stick, thinking it might be a witch. They threw the body behind the bothy and left it there…
>
> It was Malcolm MacDonald who actually laid hold of the bird, and held it by the neck with his two hands, till [the] others came up and tied its legs. It used to make a great noise like that made by a gannet, but much louder, when shutting its mouth. It opened its mouth when any one came near it. It nearly cut the rope with its bill. A storm arose, and that, together with the size of the bird and the noise it made, caused them to think it was a witch. It was killed on the third day after it was caught, and McKinnon declares they were beating it for an hour with two large stones before it was dead: he was the most frightened of all the men, and advised the killing of it. The capture took place in July. The bird was about halfway up the Stack. That side of the Stack slopes up, so that a man can fairly easily walk up. There is grass upon it, and a little soil up to the point where they found the bird. Mr Evans says that he knows there is a good ledge of rock at the

> sea-level, from which a bird might start to climb to the place. Mr Evans tried in vain to fix the *exact* year in which this event happened, but could only get 1840[4] as an approximate estimate.

Writing in the *Handbook of British Birds* (1938–41), F.C.R. Jourdain described these accounts as 'strong evidence' of the occurrence of the species at a comparatively late date. In *The Status of Birds in Britain and Ireland* (1971), produced by the records committee of the British Ornithologists' Union, this bird is declared unequivocally to have been 'the last British specimen', notwithstanding the very long period between the actual occurrence and the reporting of it.

Whether or not the account by the superstitious Laughlan McKinnon is to be preferred to that of a still later sighting reported from Ireland, perhaps the reader should decide. The highly respected William Thompson, in the third volume of his natural history of the island, gave the following account on the basis of personal communication received from the observer:

> I have little doubt that two great auks were seen in Belfast Bay on the 23rd of September 1845 by H. Bell, a wildfowl-shooter, whose good observation has already, more than once, been alluded to.[5] He saw two large birds, the size of Great Northern Divers (which were well known to him) but with much smaller wings. He imagined they might be young birds of that species until he remarked that their heads and bills were 'much more clumsy' than those of the *Colymbus.* They kept almost constantly diving, and went to an extraordinary distance each time with great rapidity.

There are no further meritorious records of the Great Auk from the waters round the British Isles. In 1846, as related in the preceding chapter, an Englishman visiting Iceland was offered two Great Auks and two eggs for the huge sum of twenty pounds. Perhaps this record has been treated with scant respect since, coupled as it is with the information that the birds went to the Zoological Museum at Copenhagen while the eggs got broken, the specimens may have been confused with those transmitted by Etatsraad Eschricht in or about 1844. Unfortunately, as it is not clear whether David Graham saw the two birds with his own eyes, there remain a number of unanswered questions about these specimens — even supposing the record to be genuine.

The party of four Great Auks observed off Vardø, East Finmark, in 1848 (also referred to in the preceding chapter) was considered to be 'an actual and unimpeachable case of the appearance of the Garefowl — perhaps of the last appearance of all,' by Professor R. Collett of Christiania (Oslo) University. This record had already been drawn to the attention of Japetus Steenstrup, while he was researching for his extensive monograph of the species, by a young scientist by the name of Nordvi (a friend of the man who observed the birds and shot one of them). Steenstrup wrote: 'I asked Mr Nordvi if a large unknown sea-bird that

Mr L. Brodtkorb had killed in 1848 near Vardø, couldn't have been of this species; but in a letter of this year [1856] he informs me that in spite of his extensive researches, no-one has ever heard of it in those northern regions.'

This gives us the reason why this record of a sighting has seldom been accorded the respect it deserves, for no sooner had Steenstrup established that the Great Auk, contrary to widespread belief, was not a bird of the high north than, immediately placing the results of the latest research in jeopardy, the species was rumoured to have been seen in the Varanger Fjord, some two hundred miles above the Arctic Circle. Nordvi, with no firsthand knowledge of the species and an academic reputation to think of, seems to have been reluctant to pick up the gauntlet which Steenstrup's enquiry represented, in spite of the compelling description of the bird given to him by his friend.

In the English-speaking world, Alfred Newton was credited with establishing the true range of the species and for a long time[6] he was almost dogmatic about its never being found above the Arctic Circle and seems to have persuaded John Wolley that, on the basis of the inadequate description of the shape of the beak, whatever the bird was that Lorenz Brodtkorb had shot, it could not have been a Great Auk. The fact remains, however, that bones of the Great Auk have been found in ancient refuse heaps on the north coast of Norway well above the Arctic Circle, giving rise to a presumption if not actual proof that they once bred in the vicinity.

In 1883, Professor Collett got to know Lorenz Brodtkorb, then aged 55, quite well. Moreover, Nordvi was by then one of Collett's colleagues at Christiania. In March 1884, at Collett's suggestion, Nordvi set down what had transpired thirty-five years earlier:

> In December 1848 I received at Mortensnäs, to the south of Vadsø, then my place of abode, a visit from my friend L. Brodtkorb of Vardø. On my asking him — who had been brought up in Vardø and was from boyhood familiar with all the birds and fishes there, and whom I knew to be an eager sportsman and good observer — what in the way of novelty he had to tell me about the animal kingdom, he told me that in the last days of April he had, in a sporting tour in the strait between Vardø and Renø, come upon four birds hitherto unknown to him, one of which he had shot and taken away with him, but had afterwards thrown away upon the shore. I asked him if the bird shot might not perhaps be one of the larger divers. He said that could not be since he had shot many birds of that genus. When he stated that the bird killed by him had no proper wings, and, as he considered, could not fly at all, because it used its wing-stump (*vinge-lapper*) to aid it in swimming, and when he mentioned, in addition, that it had a large white spot beside the eye, the thought at once came to me that this might have been the *Alca impennis*. To be surer of the matter, I asked him to look over a book containing copper-plates which I had, and to see if he could find there the bird that he had shot. Without any hesitation he pointed to the *Alca impennis* and said, 'There it is.' I then gave him some details regarding the *Alca impennis* and its history, and

> asked him to use every effort to discover if the other three birds should yet show themselves; but none of them were afterwards visible.

Three months earlier, Brodtkorb had committed his account to paper — again at Professor Collett's invitation:

> I was rowing on that day with some companions over to Renø, when we espied in the strait four large birds that attracted our notice. One of my companions, Herr Wind, now Tensmand Wessel, asked me to fire at them, in order by that means to learn exactly what sort of birds these could be which, instead of flying, only paddled upon the water with their wings. I fired, and one fell. We were all perfectly convinced we had never before seen that kind of bird. It was the size of a Brent Goose. Its back was black, and, so far as I can remember, its whole head and neck were of that colour, but in other respects it was in shape like an auk.
>
> I remember particularly that we observed a white spot at the eye on the side of the head. On the other side the ball, which had gone through the head, had torn away a piece of the white spot and shattered the beak, so that as regards the form of the beak I can tell nothing. The wings were so small that we were all agreed that this circumstance was the reason why the bird only paddled... The bird was placed in the boat in order that it might be kept; but when we reached land, it was so soaked through with water and blood that we threw it away upon the shore, though it was my intention to examine it afterwards more minutely. But when I went to get it on the following day, it had been washed away by a high sea during a storm in the night...A day or two later I was out again to seek for the remaining three birds; but I never found them. I remember, likewise, that several fishermen had taken notice of these birds before I shot the one referred to; but subsequently they were never seen again.

Professor Collett wrote back to ask Brodtkorb to inform him of any observations he might have made on the nature, voice, etc. of this bird. To this request he received the following reply:

> On the day when I shot the bird a storm was blowing from the south, so that there was a rather heavy sea. The birds were swimming right against the wind, and as we were rowing in the same direction, we got a sight of them, when they were about twenty-five yards straight in front of the boat, without flying up. In swimming they used both wings and feet, and also dived, but did not stay long under the water. It almost seemed as if they only went through the tops of the waves. The birds kept together, and did not seem afraid. We also heard a cry which they emitted when they drew more closely together. It resembled a cackling, as if they wished to call one another. At first I did not think of shooting, for the boat was rolling hard. It was only when the birds had removed to a distance of about seventy yards and were only visible at intervals, that I resolved, at the request of my companions, to take aim. When the shot went off, all the four birds disappeared; but shortly after I saw the remaining three paddling on farther until they disappeared behind the surging waves of the current.

Professor Collett concluded:

> To what is here said I need only remark, in addition, that during the winter the [Great Northern Diver] is along the coast of the whole of Finmark (and also of all the rest of Norway) a perfectly well-known bird, which is called by our sportsmen Immer or Hav-Immer (i.e. 'Sea-immer'). No confusion with it can therefore have taken place...Anyone who is at all acquainted with the nature of our different sea-birds will besides have remarked that none of the diver [*Gavia*] species keep close together when they are lying on the water, whilst this is exactly a peculiarity of the members of the auk family.

Although there are no further plausible recorded sightings of the Great Auk from European waters, after the one just described, reports of isolated examples of the species off Newfoundland were still received from time to time. In 1861, Alfred Newton (who still hoped that a 'scanty remnant' of the formerly abundant Penguins had found a safe haven from their pursuers), in drawing to a close his article in *The Ibis,* based on the late John Wolley's researches, referred to a letter received from his friend Colonel Henry Drummond-Hay:

> [I]n December 1852, in passing over the tail of the Newfoundland banks, he saw what he fully believes to have been a Great Auk. At first he thought it was a Northern Diver; but he could see the large bill and white patches,[7] which left no doubt in his mind. The bird dived within thirty or forty yards of the steamer.

Fig. 3 Four Great Auks reported off Vardø, Norway, April 1848. (Illustration by Jan Wilczur)

Alfred Newton had no need to emphasize the credentials of his friend to readers of *The Ibis*, for Henry Maurice Drummond-Hay of Seggieden in the county of Perthshire, Scotland, was a founder member of the British Ornithologists' Union, then in its third year of existence, and in time became its first President. Newton was also informed by his friend that, in 1853, a bird identified as a Great Auk had been picked up dead in Trinity Bay, Newfoundland.[8]

Finally, Henry Reeks, an early oil company executive, who wrote a sketch of Newfoundland's birdlife after his visit there in 1868, was informed by some of the settlers that:

> ...a living 'Pinwing' was caught by one Captain Stirling 'about twelve years ago', but whether destroyed or not I could not learn: Capt. Stirling was drowned and his vessel wrecked some seven or eight years since. I have no doubt that this tale is true in the main; the only questionable part being the *exact* date.

There are a number of accounts even later than this, but little to substantiate them. A native of Greenland, in either 1859 or 1867 (the accounts vary), is reputed to have shot a bird which he had never seen before but which, from its

Fig. 4 Colonel H.M. Drummond-Hay (1814–96).

description could, according to one contemporary writer, have been no other than the Great Auk. He and his companions ate it, and the dogs in his sledge got the refuse 'so that only one feather could afterwards be found'. The same fellow is reported to have said later that there were two such birds but that one had escaped among the rocks.

If we accept that the majority of Great Auks which used to visit Greenland were birds dispersing away from Iceland in the winter (and hence were very unlikely to come ashore), it is difficult to believe that any were found after the late 1830s when, according to apparently reliable information imparted to Sir Leopold McLintock, such a bird was obtained. This last reported occurrence probably refers to an incident related to Dr Edward Charlton and mentioned by him in a paper on the history of the Great Auk read to the Tyneside Naturalists' Field Club in 1859 (see Chapter 15, Footnote 3). It appears that in 1836, while in Copenhagen, he was informed by Dr Pingel that two of the species had been killed in Greenland since 1830 — one being eaten by the Moravian missionaries as a wild goose; the other preserved, ultimately, in the collection of an ornithologist at Schleswig.

Writing in 1888, F.A. Lucas of the United States National Museum stated that there had been a rumour current some twenty years previously that the Great Auk still existed in small numbers 'on the Penguin Islands in the mouth of Groswater Bay, 16 miles from Grady Harbour, a locality about 250 miles north of Cape Norman, N.F.'. Lucas judiciously concluded that although this was possible, it was hardly probable.

Notes

1 The published account of this bird in Volume III of W. Thompson's *Natural History of Ireland* (1851), although substantially accurate, included errors respecting the spelling of surnames, the roles played by various individuals in the events subsequent to the bird's capture, and the serious, if amusing, misconception that the bird habitually scratched the back of its head with its foot. The following details are based on Thompson's account as corrected by J.H. Gurney, Jnr., writing in the *Zoologist*, 2nd Series, Volume III (1868).

The bird was presented while still alive, though some considerable time after its capture, to Dr Burkitt by Mr Francis Davis who stated that it was captured a short distance from the shore, at the mouth of Waterford harbour off Ballymacaw, by the man from whom he bought it. According to its captor it was taken with little difficulty on account of its weak condition. The fisherman had kept it for some days feeding it potatoes mashed in milk. A few days later he sold the bird to Francis Davis who promptly transmitted it to Mr Jacob Goff of Horetown, Co Wexford, where it lived for about four months — although it refused food for about three weeks and had, eventually, to be force-fed. Thereafter it ate voraciously until shortly before it died. Dr Burkitt was able to supply his correspondent with a few notes reconstructed from his memoranda which create an impression strikingly consistent with that given by the Revd Dr John Fleming (see Chapter 1) regarding the bird which the latter had taken into care off Harris in 1821. 'The auk,' wrote Dr Burkitt, 'stood very erect, was a very stately looking bird, and had a habit of frequently shaking its head in a peculiar manner, more especially when any particularly favourite food was presented to it (that is if a small trout, for instance, was shown it).'

William Thompson concluded his own description of this particular bird by adding that it was reported to have been rather fierce. The bird was preserved through the intervention of an army officer, Captain John Spence, who, recognizing its interest to naturalists, while paying a social visit to Jacob Goff, 'considerately bespoke it' for Burkitt's collection, should it die. But for this recommendation, the bird's remains might have been discarded, in which case the record would, in all probability, never have come to light. Burkitt described the bird as a female in its second calendar year.

Another Mr Davis — Robert Davis of Clonmel, a noted ornithologist — received a report from the Revd Joseph Stopford in February 1844 of a Great Auk having been 'obtained on the long strand of Castle Freke (in the west of Co Cork), having been water-soaked in a storm'. Unfortunately this record is not dated and the remains were disposed of without having been adequately described, so it is impossible to state whether it refers to the same occurrence as that of the Waterford bird or another at about this time, or even whether the bird concerned was properly identified.

2 The author proceeded to identify this bird with that which escaped from John Fleming's custody on the west coast of Scotland eight years earlier.

3 Apparently this is the same Donald McQueen who described the capture, in 1821, of the bird which came into the custody of Fleming. It has never been satisfactorily explained why he made no mention of the Stac-an-Armin bird to R. Scot Skirving in June 1880 (see Chapter 1).

4 The date was fixed with reference to Laughlan McKinnon's age — he recalled having been about thirty at the time.

5 Although not an ornithologist, Bell was an unusually discerning wildfowler who took particular notice of the occurrence of any unfamiliar species. He is credited with the first American Wigeon and the first Broad-billed Sandpiper on the List of Birds recorded in the island of Ireland.

6 Later, in the early 1880s, Alfred Newton retracted a footnote he had inserted in his paper in *The Ibis* (1861) calling into question the authenticity of a

record of 1815 from Disko Island, lying just north of the Arctic Circle, on the west coast of Greenland.

7 It is not clear if we should understand the bird to have been in breeding plumage or simply a well-marked specimen in winter plumage.

8 Per J. MacGregor, *in litt.*

CHAPTER 14

The natural history of the Great Auk

In summarizing the little that is known of the breeding habits of the Great Auk, the nineteenth-century ornithologist C.E. Dixon wrote: 'Unfortunately it ceased to exist before the era dawned when the habits of birds were studied minutely and in careful detail, so that our information is of a general character only.' Accordingly, to reconstruct the Great Auk's natural history from the sparse historical records, one must draw on the natural history of its surviving relatives — particularly the closely studied Razorbill and Guillemot — and by establishing what features of courtship and breeding they have in common, to infer what intimacies were shared on the Great Auk's remote breeding grounds.

While many of the Auks, and indeed other sea birds, reveal a remarkable degree of convergence in certain aspects of their courtship and reproduction, each alcine species has evolved other habits according to its particular mode of life. Importantly, the periods of both incubation of the eggs and care of the young at the nest site vary markedly as each endeavours to maximize the survival rate of its offspring. The burrow-nesting Puffin, for instance, tends its single chick below ground for seven weeks; the crevice-nesting Razorbill and cliff-nesting Guillemot lead their progeny off to sea when half-grown; while three species of Murrelet in the Pacific (Craveri's, Xantus', and the endangered Japanese Murrelet) brood their young for less than forty-eight hours — until, that is, the internal thermostatic regulator of the newly hatched chick has developed sufficiently for it to take to sea without perishing of hypothermia. The last-mentioned young are usually known as 'semi-praecocial' since they are nearly, but not quite, as 'praecocial' as the nidifugous young of gamebirds, waders, and wildfowl which quit the nest as soon as all the eggs are hatched and they themselves are dry. The young of a fourth species, the Ancient Murrelet, although it does not go to sea immediately, is capable of thermo-regulation on hatching and so may be said, broadly speaking, to be praecocial.

Whether the Great Auk, unlike any of its close surviving relatives, hatched a praecocial chick is a question it is no longer possible to answer decisively. John Gould stated unequivocally in his *Birds of Europe* (1837) that 'the young take to the water immediately after exclusion from the egg and follow the adults with fearless confidence'.

It is unfortunate that Gould cited no authority for this assertion. Ornithologists today wonder if anyone in years gone by was privileged enough actually to witness the hatching of young Great Auks, or whether the behaviour described by Gould was simply a matter of inference. Nevertheless, as it is doubtful if such an assertion would ever have been made in the absence of at least circumstantial evidence, it would not seem unreasonable to assume that recently hatched young in the company of parent birds were observed from time to time. As we shall see presently, documentary evidence from the seventeenth century corroborating Gould's assertion has indeed come to light.

Quite apart from Gould's testimony, there are a number of indicators suggesting that the Great Auk did indeed hatch highly praecocial young. For instance, the size of an egg stands in relation not only to the size of the adult bird but also, in general, to the degree of praecocity of the young on hatching. A bird which lays a relatively small egg may be expected to have only a brief incubation period. An extreme example is the European Cuckoo which lays a tiny egg for its size, the young hatching blind after a mere twelve and a half days. By contrast the Whooper Swan, with an egg averaging just under 4.5 inches (11.4 cms) in length, incubates for five to six weeks, while the Great Northern Diver, with an egg of just over 3.5 inches (8.9 cms) long, incubates for a month. The downy young of both of these species are brooded for some time after hatching. The egg of the Great Auk, with an average length of nearly 4.9 inches (12.4 cms) would appear, prima facie at least, to have required a proportionately longer incubation period, thereby increasing the likelihood that the young would emerge endowed with a high degree of praecocity.

However, in addition to the actual length of an egg, its size in relation to the weight of the parent bird is an important factor, clearly illustrated in the case of the small Pacific Auks that do hatch semi-praecocial young. Calculations made by Professor T.R. Birkhead to discover if the egg of the Great Auk was also disproportionately large indicated that it was in fact a degree smaller relative to the size of the adult than that of the Razorbill.

Martin Martin's testimony, derived from the St Kildans with whom he conversed in June 1697, was that the Great Auk was seen only for six or seven weeks of the year. Rather than assuming this to be a period for incubation, we should regard these six or seven weeks as including the time necessary for courtship and ovipositing. Moreover, the period is probably longer than that for which any single bird came on shore as it almost certainly covers the time between the first and last annual sightings of the species. Consequently, unless the Great Auk had an unusually short incubation period, there would seem to have remained precious little time for a bird laying so large an egg to tend its young at the nesting site.

During his visit to Iceland with Alfred Newton in 1858, John Wolley gathered valuable information on the duration of the Geirfugl's breeding season on Eldey from Vilhjálmur Hákonarsson, who had led the expedition which had obtained the last Great Auk specimens. In his *Garefowl Book* Wolley wrote the following entry for the 28th of May of that year:

> Vilhjalmur tells me again that the Garefowl lays eggs first about 29th & 30th of May — first young come out about Midsummers day, as Brandr [Brandur Guðmundsson, 1771–1845, the 'formaður' of the fowling expeditions in the early 1830s — see Chapter 12] experienced once in finding young large in the eggs about that time — So the bird is altogether about five weeks on the rock he says, & till the 6th of July — Once, Vilhjalmur says, when Brandr went on the Fardagar 3, 4th & 5th of June there were fresh eggs — It is perhaps a *little* earlier than the Svartfugle–Guillemot. Obs. Though Vilhjalmur cannot give the clearest proofs of each of his opinions, it is evident from the care with which he speaks with the Almanak before him, that he speaks to the best of his ability, and that that knowledge of his is founded on a consideration of tradition & of all the facts that have come to his notice, all which he as leader of the later parties has been obliged to keep in mind

This testimony, which has hitherto gone unnoticed, is valuable in that it clearly indicates an incubation period of approximately 26 days — a remarkably brief period for so large a bird. Four days earlier Vilhjálmur Hákonarsson had told John Wolley that it was 'not long before the young leave the rock — perhaps twelve or fourteen days'. This information, however, does not appear to have been based on personal observation and Wolley, regarding it as conjecture only, commented that Hákonarsson had 'no good reason as it seems to me' for holding such an opinion. The phrase 'until the 6th of July', accordingly, should be taken as no more than a computation based on the assumption that, like the other cliff-nesting auks, the Geirfugl took its young to sea when half-grown. While it is not impossible that the young were tended on the rock, it is remarkable that none of the inhabitants of the Reykjanes peninsula, including Vilhjálmur Hákonarsson himself, ever mentioned having seen juvenile Great Auks. Moreover, it will be recalled that on 1st July 1821, Friedrich Faber failed to find any Great Auks at all at the Geirfuglasker. These facts, when taken together with the relative brevity of the incubation period, suggest that the Great Auk had evolved a breeding strategy that required it to remain ashore only for the shortest conceivable period.

One of the first ornithologists to address the question of the relationship between the Great Auk's natural history and its survival strategy was the youthful Karl Micahelles (1807–34). The following conjecture, published in the journal *Isis* in 1833, in implying an extensive period of care of the chick at the breeding place, is based on the presumption that the Great Auk, the Razorbill, and the Common Guillemot shared the same habits in bringing up their young:

> The peculiarity previously raised in relation to the St Kilda population that [the Great Auk] never lays an egg a second time if the first be taken away, can perhaps stand in some kind of relation to the dissimilar mode of feeding the young and the great difficulty that the bird, under certain conditions, must have with the youngster's food if he can only supply the latter by swimming and on foot, whereas his congeners can bring the food to the nest on the wing.

Micahelles' questionable inference does serve, indirectly, to emphasize two entirely separate points — first and foremost, the compelling need for the Great Auk to get its young off to sea quickly; and, secondly, the inestimable value which attaches to Martin Martin's laconic account of the species' natural history (given in Chapter 7).

As has already been intimated, information tending to corroborate the theory that the Great Auk's young was praecocial — or at the very least semi-praecocial — on hatching may be found in the writings of two other seventeenth-century[1] writers, Nicolas Denys of France and the Plymouth surgeon, James Yonge. Concerning the '*Pennegoin*' Denys wrote:

> [I]t does not fly, having nothing but two stumps of wings with which it beats against the water to aid it in fleeing or diving. It is claimed that it dives even to the bottom to seek its prey upon the Bank. It is found more than a hundred leagues from land, where, nevertheless, it comes to lay its eggs [*sic*] like the others. When they have had their young they plunge into the water; and their young place themselves upon their backs, and are carried like this as far as the Bank. There one sees some no larger than chickens[2] although they grow as large as geese.

A decade earlier, in 1663, Yonge observed what he called 'strange coloured gulls, Penguins; a bird with a great bill and no wings but such as goslings have. They can not fly, but when pursued, take their young on their back.'

These are the only known contemporary references from the New World to the Great Auk's habit of carrying its young either to get out of the way of shipping or to ferry the chick to suitable feeding grounds. The time spent nestling in the soft plumage on the back of the parent bird while covering the eighty-five miles from Funk Island to the nearest point on the Banks, might have been a sufficient substitute for any need for initial brooding that the newly hatched bird might otherwise have had.

In order to lay an egg large enough that the chick might speedily take to the safety of the sea, the Great Auk needed to be a bird of considerable size. As there is, besides, a relationship between size and flightlessness, we may observe a number of factors each operating independently but together tending to point to an optimum size for the species. As the anatomist F.A. Lucas remarked in

Fig. 1 A Great Auk carrying her chick. (Illustration by Jan Wilczur)

respect of the huge flightless birds of the Southern Hemisphere, such as the Moa of New Zealand, 'once established, flightlessness and size play into one another's hands' — a comment which may be applied by analogy to the Great Auk. As Lucas pointed out: '[T]he flightless bird has no limit placed on its size while granted a food supply and immunity from man.' In the case of the Great Auk, food supply, or to be more precise, the opportunity to exploit a food source denied to smaller species, was a factor which made increased size a favourable evolutionary path for the species to follow. The inability to fly, we may reasonably conjecture, came about when a large population of Auks — or, perhaps, 'proto-auks' — lost the habit of flight because of the immediate proximity of an abundant food supply near low rocks at the water's edge.

The larger the diving bird, the deeper it can go in search of prey, and it is significant that two species of fish — the Shorthorn Sculpin of the genus *Myoxocephalus* (Bullhead) and the Lumpfish (or Lumpsucker) *Cyclopterus lumpus* — which, besides crustaceans, constituted some of the Great Auk's principal dietary items, are both bottom-dwelling kinds.[3] The limiting factor which prevented the Great Auk from becoming a giant among waterfowl, as the prehistoric *Hesperornis* had done, was the potential loss of manoeuvrability and speed in deep water which would have resulted from a much greater size. The depth of the water on the Great Bank, it will be recalled, is from 15–60 fathoms

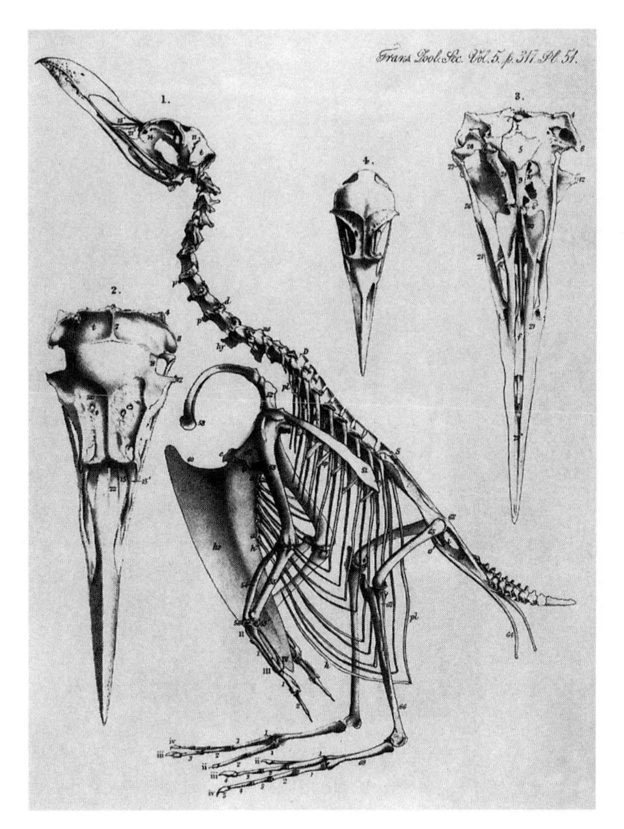

Fig. 2 Lithograph by E.A. Smith from Professor Richard Owen's *Description of the Skeleton of the Great Auk or Garefowl* (1865).

(90–360 feet or 30–120 metres) and in the channel between the Bank and the island of Newfoundland, where the Great Auk was frequently encountered, the depth is even greater. It seems probable that the Great Auk was quite capable of feeding at depths considerably in excess of those which human beings are able to endure. Indeed it is not beyond the bounds of possibility that the Great Auk may have been able to rival the Emperor Penguin of Antarctica which has been recorded at the phenomenal depth of 870 feet (approximately 235 metres), for even the much smaller Common Guillemot has been reliably reported, on occasion, as having been trapped in fishing nets as deep as 180 metres.

In 1885, in a letter to H.W. Feilden, who had undertaken research into the Great Auk's former status in the Faeroe Islands (see Chapter 11), Alfred Newton wrote from Cambridge:

> I can't *satisfy* myself as to the way in which the Garefowl's flightlessness was produced, and I suppose I never shall. I can only conjecture that he found wings fit for flight articles too expensive for him to indulge in. If he descended from a

> Razorbill, it is not difficult to imagine that he found big wings were not worth the trouble of growing, and it was better to expend energy in simply accumulating bulk. But one has no more right to assume that he descended from a Razorbill than that the Razorbill descended from him. The most reasonable conjecture seems to be that they had a common ancestor, who differed in some degree from both, but still one would think that common ancestor must have had the power of flight. Such natural enemies as that common ancestor (or the Razorbill for the matter of that) possessed may be roughly divided into two categories: enemies in the air or on land, and enemies in the water. Now in the water, wings, to an alcine bird, are chiefly useful for steering (the propelling power being in the leg) and a very little bit of wing would do to steer with, and escape from a grampus or seal. In the air a wing must be very good to be good for anything, if not it is better not to fly at all...Natural selection would soon weed out animals with moderate wings and leave those which had the best or the worst. On land I take it the Garefowl had practically no enemies till man came to civilise him.
>
> I don't say these views satisfy me, there may be considerations I have altogether overlooked, but I think they may serve as indications of something like the way it was done.

It is now a generally acknowledged fact — and therefore difficult to understand how Alfred Newton appeared not to recognize it — that the Auks habitually use their wings for underwater propulsion, and not their feet only as is the case with the divers. Colonel Montagu, it may be recalled from Chapter 1, had described the wings as acting 'as fins' when the Great Auk was in pursuit of its prey. The greater the wing area, the greater the resistance it gives to the water. As a bird which took its prey from on, or near, the seabed, the Great Auk would have been seriously disadvantaged if its wings had been in proportion to the increased bulk that it was obliged to acquire to enable it to reach the depths to which it aspired.

Every aspect of the Great Auk's structure marks the species out as having achieved a high degree of perfection as a specialized feeder in deep water. Were it now possible, with the aid of a small underwater craft, for biologists to observe Great Auks catching lumpfish off the Newfoundland Banks, the birds would appear considerably smaller and more compressed on account of the intense water pressure, than they would when on the surface or on land, for the bird's skeletal structure was characterized by both strength and flexibility.

The greatest threat for diving birds is to the vital organs, especially the heart and lungs, from the intensity of the pressure at any considerable depth. Eight V-shaped ribs, fitted with an articulated joint where they met the main breast bone (sternum), and overlaid with a fine 'basket weave' of muscles and anchored to a rigid spine, the central vertebrae of which were fused, gave the Great Auk the protection it required. The broad and somewhat flattened bones of the 'forearm' of the wing — particularly the radius and ulna — in the words of the comparative anatomist Professor Richard Owen, who in 1865 published a paper on the skeleton of a Great Auk brought back from Funk Island, 'relate to the support of a surface in the shortened wing adequate, as a fin, to strike the water with effect'. The pectoral muscles, which lay either side of the massive sternum, gave the bird the requisite power to chase its prey over some distance. The powerful legs and feet, set far back as in the case of grebes and divers but with the added combination of short tarsi below the leg joint and relatively long shanks (or tibiae) above, served to give the species added thrust as it closed in on its intended victim. Finally, the supple neck, capable of retraction and extension, lent the Great Auk greater manoeuvrability as it flew, quite literally, through the water, as well as the capacity to lunge forward at the last moment of the chase.

A good many of the external features of the Great Auk also reveal their adaptation for underwater pursuit. To prevent water entering through the nostrils and causing brain damage, not only were the apertures the merest slits along the base of the upper mandible, as they are on the Great Auk's relative, the Puffin, but they were protected by an inner membrane and dense, soft feathering too. The nostril cavity in the bone structure of the beak, beneath the horny covering

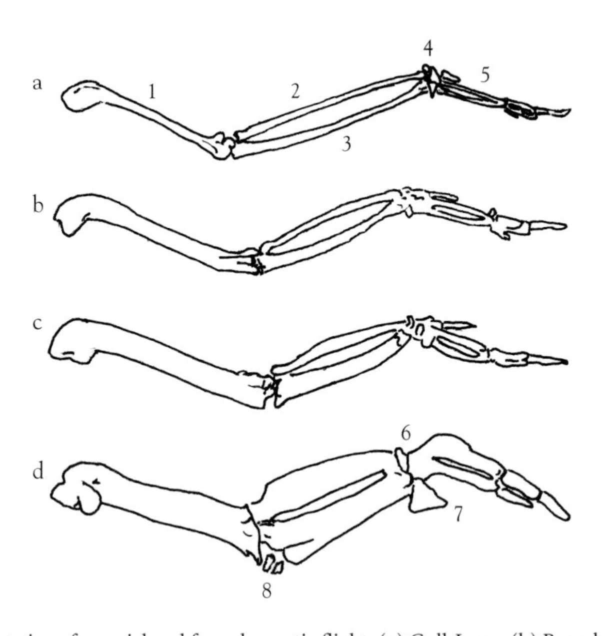

Fig. 3 Adaptations for aerial and for subaquatic flight: (a) Gull *Larus*; (b) Razorbill *Alca torda*; (c) Great Auk *Pinguinus impennis;* and (d) Penguin *Spheniscus.* (1) humerus; (2) radius; (3) ulna; (4) carpal joint; (5) carpometacarpus; (6) scapholunare; (7) pisoulunare; and (8) sesamoid bones. The bone structure of *Spheniscus* is much more developed for underwater flight than that of *Pinguinus.*

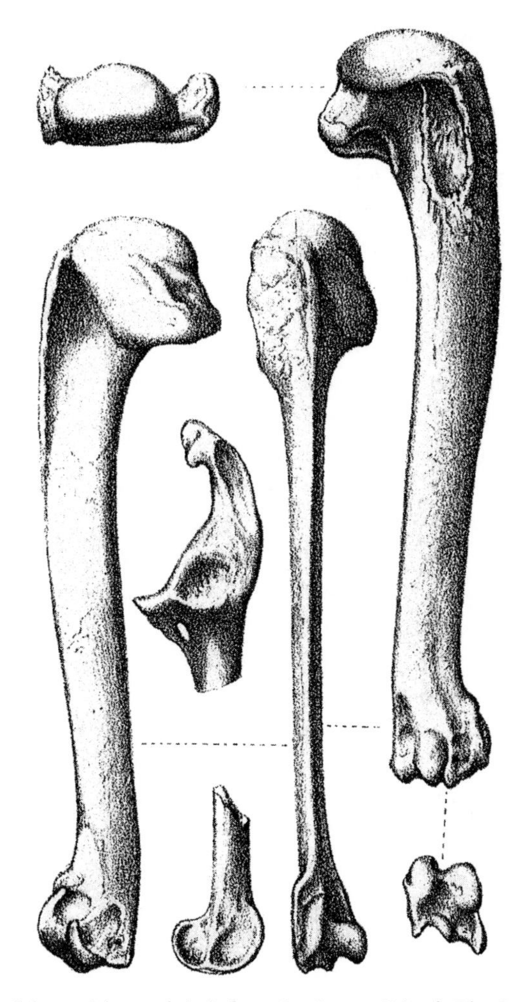

Fig. 4 Great Auk humeri (natural size); from Symington Grieve's *The Great Auk or Garefowl* (1885).

(or rhampothecus) common to all birds, was, in contrast, massive. This enabled the bird both to inhale deeply before diving and to exhale by small degrees almost continuously whilst in the depths, and thus offer resistance to the water which would otherwise have found its way along the nasal passages to the brain cavity. To prevent water gaining access via the ear, this too was proportionately very small, although the bird's powers of hearing, according to information given to John Wolley and Alfred Newton, remained acute. The eyes, too, were small for a bird of the Great Auk's size. Even the texture of the plumage, likened by some of the older writers to silk, was a last layer of protection against the rigour of its chosen environment.

The majority of diving birds, including a number of freshwater ducks, are strongly patterned with black and white, a feature which has considerable advantages when fish are being pursued, for the intended prey is denied a clear view of an approaching silhouette. We may term this, perhaps, 'offensive camouflage'. A variegated head pattern is no more than a refinement of this principle and one which is by no means confined to birds. An illustration of 'convergent evolution' is found in the large, oval, white spot on an otherwise dark head on the Gentoo Penguin — as we know, no relation of the Great Auk — and also on the Killer Whale. As only a greatly reduced amount of light filtered down to the depths at which the Great Auk hunted, we may imagine that its prey would have been easily confused as to the immediate proximity of danger, for the white spot would have diverted attention away from the dangerous bill.

The size and strength of the bill lend weight to the theory that the Great Auk regularly took crustaceans, such as crabs of some size, which require both a firm hold and considerable power to crush them once secured. An anecdote from Iceland, recorded by F. Benicken in the *Isis* for 1824, well illustrates the formidable power that the Great Auk possessed in its strong lower jaw. A young lad, who landed with a fowling party one year on the Geirfuglasker, was bitten so severely on the forearm by a cornered Garefowl that, in spite of a calfskin jacket that he was wearing at the time, his blood ran freely down his sleeve. We should note that, despite the large size of the Great Auk's bill when viewed from the side, it was laterally strongly compressed, making it powerful but streamlined. In addition to its narrow width, the bill was grooved after the fashion of the colourful rhampothecus of the Puffin, enabling water to channel off rapidly.

As the spot between the beak and the eye was intermixed with blackish feathers after the autumn moult, we may infer that the gleaming white spot on the adult bird in breeding plumage was significant for courtship and the selection of a mate. Although the record is silent on this aspect of the Great Auk's natural history, we may, as stated in the opening paragraph, reasonably deduce that where particular types of behaviour are shared by two or more closely related species, or extend across a range of species not all of the same family, something similar would have been found in the Great Auk's nuptial display. What we can be certain of is that the Great Auk's breeding grounds would have been pulsating with procreative energy. That experienced observer of sea birds, Richard Perry, gave a lucid account of the cliff life of Auks (including the Great Auk's closest surviving relative, the Razorbill) in his *Lundy, Isle of Puffins* (1940). Of the pair-bonding between individuals he wrote:

> The Razorbill is obviously a species whose impulses are strong and easily stimulated, so that it is perhaps not very remarkable to find him slipping so facilely from one impulse to another, knowing as we do that a bird is a creature in whom every organ works in conjunction with another, and whose every action is a result of the interplay of a number of different internal and external stimuli.

> With that inconsequence common to wild birds, a razorbill fighting furiously with a rival, will switch in a moment to affectionately nibbling his mate. The Razorbill and the Guillemot impressed me most, I think, by their extraordinary and most unbirdlike responsiveness to physical sensations. There was no confusion on this point: some things were pleasurable to them, and some were distasteful, and they left one in no doubt as to which were which.
>
> Before his mate the male Razorbill occasionally bends his head right back over his mantle, in the manner of a cormorant, with a slight opening of that gamboge buccal cavity[4] that is in such artistic contrast to his black and silver plumage. He follows this antic by strongly pecking his indifferent partner's averted head, eventually provoking her to further endearments which eventually take the form of mutual preening. This preening of one another's feathers is so dominant a feature of the cliff life of razorbills, and especially of guillemots, that one wonders whether it was originally, and perhaps still is, a utilitarian device for removing dirt and lice…
>
> The laying back of the head is a rather rare razorbill antic. Most frequently the male (and also the female) elevates his head and bill vertically, his slightly parted mandibles vibrating swiftly with the rattle of castanets, while his mate nibbles his taut throat, and he shivers in an ecstasy of pleasurable sensation: for this antic is usually associated with extreme physical pleasure or excitement.

In 1887, soon after his return from Funk Island in search of Great Auk bones for the Smithsonian Institution, F.A. Lucas wrote bleakly of the demise of the species: 'The naturalist can but regret its wanton destruction and deplore the loss of so interesting a bird.' As an anatomist, Lucas was thinking primarily of the Great Auk's unique status as the Northern Hemisphere's only flightless diving bird. Intimate descriptions of other Auks, such as that already given, can only redouble the sense of what has been lost and add greater poignancy to his words.

> There is something in the strange figure and aspect of the Penguin well agreeing with the wild lonely remote islands in which it congregates. In beholding a spot on the surface of our globe, ocean-girt and uninhabited by Man, tenanted by thousands of these birds, which for ages — generation after generation — have been in uninterrupted possession of the place, we are thrown back upon primeval days, and we involuntarily recur to the now extinct Dodo, a wingless bird, which formerly tenanted the islands of Bourbon, Mauritius, and Rodrigue, once desolate and untrodden by the foot of Man, as are still many of the haunts of the Penguin, and the idea forces itself upon us that like the Dodo, this species also may at some future time become utterly annihilated.

So wrote the Newfoundland writer and historian, the Revd Philip Tocque, earlier in the same century just as, unknown to him at the time, the species was passing into oblivion.

There is evidence that the great sea bird breeding colonies off Newfoundland had been prospected with a view to economic exploitation even before colonists went out to settle on the island. The first account of Sir Humphrey Gilbert's voyage in 1583 on behalf of his sovereign, Elizabeth I, as stated in Chapter 5, was published the same year by a supporter of the expansion of English power into the New World, Sir George Peckham. Drawing attention, in his *True Reporte* of the voyage, to the abundance of wildlife on the island, presumably with the intention of encouraging settlement in the near future, this writer listed a dozen or so of the more distinctive species of birds; to this list he added: 'feathers of sundry sorts as for pleasure and filling of featherbeds'.

It was not, however, the decision to exploit sea birds for their feathers, any more than for their eggs, which, *of itself*, led to the extirpation of the species whose history is told in this book: the explanation lies in the decision of a very few people to encamp on the principal breeding ground of the Great Auk for weeks on end, in order to prosecute their (by then) illicit enterprise more efficiently. This ensured the rapid disappearance of a species which could not, being flightless, retreat to inaccessible sites. We can only account for the annual visitations of these men when we recall the enormous profits which they could expect from dealers supplying the makers of pillows and other bedding in the United States. Had the potential profits from the trade not been so inordinate it is doubtful whether anyone exploiting eggs or birds would have elected to remain on Funk Island for more than a day or two at a time. In that case, the Great Auk would, in all probability, have survived — albeit in sorely depleted numbers.

The noise, the stench, the alternation of blistering heat and sudden gales, together with the attentions of parasitic mites, and, of course, the absence of even the rudiments of civilized life, meant that a stay of several weeks' duration on Funk Island would have been more difficult to endure than a prison sentence of equivalent length. It is now five hundred years since John Cabot discovered Terra Nova; for three hundred and fifty years of that period, the Penguin (as it was called) was extant in the waters off Newfoundland's coasts. In view of the continual subjection to raids that it had to endure whenever it came ashore to breed, and the consequent decimation of its numbers, this fact alone is testimony to the Great Auk's remarkable resilience as a species.

Notes

1 Neither of these two authors was published in English until the twentieth century, although a German rendering of the relevant part of Denys' *The Description and Natural History of the Coast of North America (Acadia)* of 1672 (which had first appeared in Dutch in an early eighteenth-century economic digest devoted to the fisheries of north-eastern North America)

was cited in a footnote in Professor Japetus Steenstrup's monograph devoted to the Great Auk (1857). The *Journal of James Yonge, 1647–1721* remained unpublished until 1963.

2 Significantly, the term used in the German translation (1723) is 'kuchlein' which indicates newly hatched chicks rather than full-grown domestic fowl.

3 In his *Arctic Zoology* (1784–7), Thomas Pennant described this species as particularly abundant off Newfoundland. Other species which may also have been taken by the Great Auk include Capelin (*Mallotus villosus*), Shad (*Alosa sp.*), Three-spined Stickleback (*Gasterosteus aculeatus*), and species of the genera *Bevoortia, Gadida, Morone,* and *Teleosteus.* There is uncorroborated evidence to suggest that the young may have either taken or actually been fed *euphausiid* plankton. See further: Olson, S.L. *et al.* (1979) *Auk,* **96**: 790–2 and Hobson, K. A. and Montevecchi, W. A. (1991) *Oecologia,* **87**:528–31.

4 Richard Perry stated that raising the head vertically with parted mandibles to reveal the yellow gape is 'a prominent feature in the sexual behaviour of puffins, razorbills, kittiwakes, and all the other gulls, shags and cormorants'.

CHAPTER 15

Bird protection: a pressing need

On 23 April 1845, *An Act for the Protection of the Breeding of Wild Fowl in this Colony*, which had been passed by the Newfoundland legislature some months previously, took effect on receiving the royal assent. It thereby came into being almost sixty years after Rear-Admiral John Elliot had issued the first proclamation intended for the safeguarding of the sea bird breeding colonies, not least those on Funk Island. Even leaving to one side this and subsequent proclamations of Newfoundland's Governors, the draconian Act of 1845 constitutes the first conservation legislation of its kind in the British Empire and it would in time prove an invaluable point of reference to those who, in the decades following John Wolley's urgent recommendation to the British Association for the Advancement of Science in 1849, pressed for statutory bird protection to be introduced in Great Britain.

Writing in 1874, John Milne, the first visitor to Funk Island since Peter Stuvitz (who a generation earlier had gone for the purpose of collecting bones) commented that the trade in sea birds' feathers 'has now very rightly been prohibited for fear it should end in the extermination of "the golden goose" — a lesson learnt by the annihilation of a former inhabitant of the same spot, the Great Auk'. As related in Chapter 3, a warning in not dissimilar terms was made by J.J. Audubon after he had witnessed fowlers denude an entire Black Guillemot colony of both birds and eggs.

Regarding wildlife, we should not expect to find anything on the part of legislators approaching what in the nineteenth century was deemed 'sentimentality' any more than we would look for it among the Newfoundland fishermen and their aiders and abettors, the eggs' and feathers' merchants resident in St John's and Halifax, Nova Scotia. The prevailing attitude among parliamentarians and others who influenced public opinion was thoroughly utilitarian: any species, particularly of birds, was judged according to its usefulness to mankind; other considerations, such as vulnerability to collectors, being regarded as beyond the proper concern of the law. A different kind of utilitarianism permeated the thinking of those whose actions destroyed the Great Auk on its principal

ANNO OCTAVO

VICTORIÆ REGINÆ

CAP. XIII.

An ACT for the protection of the Breeding of Wild Fowl in this Colony.

[Passed 23rd April, 1845.]

WHEREAS it is expedient to make provision for protecting the breed of Wild Fowl in this Colony. Preamble.

I.- Be it therefore enacted, by the Governor, Council, and Assembly, in Legislative Session convened, that from and after the passing of this Act, it shall not be lawful for any person or persons, within this Island or any of its dependencies, wilfully to break, destroy, take, carry away, use, purchase, barter, sell, or expose to sell, or knowingly have in his, her, or their possession, any of the Eggs of any of the various species of Wild Fowl which resort to or frequent the shores, coasts or other parts of the Island, or of the Islands or Dependencies within the Government thereof; or by any ways or means whatsoever wilfully to remove or destroy any of the said Wild Fowl during the breeding season, that is to say, between the Tenth day of May and the First day of September, in each year: And that if any person or persons shall, after the time aforesaid, within this Island or any of its dependencies, wilfully break, destroy, take, carry away, use, purchase, barter, sell, or expose to sale, or knowingly have in his, her, or their possession, any of such Eggs as aforesaid, or shall, during the breeding season aforesaid, by any ways or means whatever, wilfully remove or destroy any of the said Wild Fowl, every such person or persons shall, for every such offence, not only forfeit the same, but forfeit and pay to Her Majesty, her Heirs and Successors, a sum not more than *Twenty Pounds*.

Trade in Eggs of Wild Fowl prohibited

Wild Fowl not to be destroyed during certain seasons

II.- And be it further enacted, that all fines and forfeitures incurred under this Act, shall and may be sued for on the oath of one or more credible witness or witnesses, or by the confession of the party, in a summary way, in any of the Superior Courts or any of the Courts of General Quarter Sessions of the Peace in this Island, and shall be levied with costs on the goods and chattels of the offender, and shall be paid into the hands of the Colonial Treasurer, for the general purposes of the colony; Provided always, that all informations and prosecutions for any of the above offences shall be made and commenced within three months after any such fine or forfeiture shall have been incurred.

Recovery of penalties

Proviso.

Printed by RYAN & WITHERS, Printers to the Queen's Most Excellent Majesty.

Fig. 1 An Act for the Protection of the Breeding of Wild Fowl in this Colony, 1845.

breeding grounds, for although the Newfoundland Penguins were the most easily exploited, there were other species of Auk and the Gannets which could, with little more effort, have been killed in their place. In the words of F.A. Lucas, who followed in the footsteps of John Milne in 1887: '[A]fter the extermination of the Great Auk the fishermen and eggers seem to have done their best to extirpate the remaining denizens of this isolated spot.'[1]

Incidental to the part played by market forces in bringing about the Great Auk's extinction, there is the question of unbridled competition. In contrast to the Faeroe Islands and St Kilda, where man and bird lived almost cheek by jowl with each other and where exploitation of the latter could be carried on co-operatively, the sea bird breeding stations off Newfoundland were uninhabited. In the former, it was in no one's interest to outdo his neighbour, but rather to ensure that he received his fair share of nature's bounty. By contrast, even if the 'poor inhabitants of Fogo Island' (to use Captain Cartwright's phrase) took only a moderate quantity of birds or their eggs, the breeding stock would not thereby be safeguarded since any accessible birds left behind would be sure to be killed by the very next crew to set foot on the place.

There is, too, one other factor at work in the extermination of New World species, and one which, paradoxically, might be pleaded on behalf of the fowlers by way of mitigation — namely, the superabundance of wildfowl which the first generations of settlers encountered. This must have encouraged immoderate destruction, for no amount of slaughter seemed, initially at least, to diminish the breeding populations of the exploited species. Such a habit, once acquired and passed down through generations, would have become difficult to shed even when the severe decline[2] of a bird — such as the Great Auk — became self-evident for anyone with eyes to see.

With a rapidly increasing population both in Europe and North America in the wake of the Industrial Revolution, the pressures on wildlife, whether through habitat destruction (as in the case of the extinct Passenger Pigeon) or the practices of the amateur sportsman (as in the case of the Eskimo Curlew), reached intolerable levels. The declining strength of sea bird colonies was one of the more visible symptoms of a situation which became more critical year by year. But the decline of 'useful' species, especially destroyers of insect pests in Western Europe, first prompted the Governments of France and Germany to consider appropriate legislation in the 1860s. Not surprisingly there were also champions of such species in Great Britain, and in 1862 the *Norfolk Chronicle* published a lengthy leading article denouncing the 'undoubted crime' of employing poisoned seed dressings to reduce the populations of what were then called granivorous birds, such as finches. The destruction was in response to a supposed population explosion among such birds resulting from the wide-spread reduction of birds of prey by those with sporting interests. Those who could give a balanced view of the place of raptorial species in the 'economy of nature' were few, and their capacity to influence public opinion severely limited. Although *The Times* regularly published letters by the well-known Yorkshire naturalist, author, and clergyman, the Revd F.O. Morris, on such matters as the scarcity of summer migrants, if whole populations of certain species were to be saved there was certainly a race against the clock.

The two decades prior to the first statute for the protection of wild birds in the British Isles belong to the tiny number of individuals clear-sighted enough to see that there was a pressing need for modern man either to exercise self control or to be controlled by law in his exploitation of birds on their breeding grounds. Although, as outlined by John Wolley, there was an urgency to protect the sea bird breeding colonies on British coasts, the Act for the Preservation of Sea Birds of 1869 should be considered as a first step in achieving comprehensive protection for wild birds during the breeding season or close-time. The ornithologists who made up the Close-time Committee (a body which could give expert evidence to Parliament) comprised that small circle of aficionados who were aware of the salient facts of the extirpation of the Great Auk — men who gradually came to realize that they had reason to lament the passing of a species for which the British Isles had represented the easterly limit of its breeding range.

In 1859, the year after Alfred Newton and John Wolley returned from their fruitless search for the Great Auk in Iceland and the year of John Wolley's untimely death, the Revd H.B. Tristram, another ornithologist who had taken holy orders and was, nonetheless, an avid collector of eggs and specimens, addressed a meeting of the Tyneside Naturalists' Field Club. During a wide-ranging speech which touched on matters as diverse as Darwinism and the demise of the Great Auk,[3] he made the following observations:

> It would really seem that, on some points, our boasted civilisation has become rather barbaric. Our moors and woods are infested with a class of men called gamekeepers who look upon the multiplication of pheasants as the first duty of man, and the destruction of every other bird bigger than a thrush as the highest effort of human genius. Too many proprietors are content to leave the management of their preserves entirely to their keepers, and while the man believes hedgehogs suck cows, and destroys them accordingly, the master is equally credulous as to the hares devoured by white owls and the pheasants carried off to kestrels' nests.
>
> In noble contrast to the careless indifference of too many landowners, I should ill discharge a debt which we all, as naturalists, owe did I not gratefully acknowledge the persevering care with which the then Archdeacon Thorp has, for several years, protected the sea-fowl of the Farne[4] Islands. For no advantage to himself, and at no little cost, the Archdeacon has taken those islands and maintained guardians (not gamekeepers) during the spring and summer months, until the young birds have been hatched and flown. The results are patent. The terns which, when he first took the rocks, were dwindling to a few dozen are already recovering their numbers; the guillemots maintain their strength; the rare Roseate Tern still lingers among its fellows; the Eider Duck, reduced to two pair, now breeds in scores, and the little Dotterel [viz. the Ringed Plover] runs along every scrap of shingle. Meantime the Eider has been chased from all its other haunts on the coast, and I

> believe that not more than a pair can be found on Coquet. St Cuthbert's Duck is a bird peculiarly interesting in this locality, not only from its traditional association,[5] but from those rocks being the most southern limit of this Arctic bird. And yet persons calling themselves gentlemen can be found who will lie in a boat all day, just out of reach of the guards, and amuse themselves by shooting the birds from their nests. Had it not been for the public spirit of Archdeacon Thorp, we should by this time probably have had to lament the Eider, the Guillemot, the Roseate, Sandwich, and Arctic Terns as no longer residents on our coast.

No spot on the British coast which lay within close reach of a railway station — if it was a sea bird breeding place of any size — was immune from the attentions of the weekend sportsman. In 1850, writing a footnote to John Wolley's paper on the Faeroe Islands in *Contributions to Ornithology*, the journal's editor, Sir William Jardine, in approving John Wolley's call for legislative action, described in strongly deprecatory language what he had observed at the famous gannetry on the Bass Rock at the mouth of the Firth of Forth:

> In the summer of 1848, the last time we visited it, two parties in boats at about a quarter of a mile distant, were shooting the [Solan] Geese in mere wantonness, allowing them to float unlifted. It is well known that they approach any boat which comes near their haunts and that when one is killed, numbers assemble and hover over it, and this habit is taken advantage of by those would-be or rookery sportsmen. The Bass used to be a favourite place for us to visit from 1816–1820; and on revisiting it in 1848, the sea-fowl had very visibly decreased.

Fig. 2 Canon H.B. Tristram (1822–1906).

> We cordially agree with Mr Wolley's suggestion for their preservation during the breeding season.

The Scottish naturalist, John Rennie, who was Professor of Zoology at University College in London, has left a description of a visit to the island of Berneray in the Outer Hebrides which gives an impression of what the sea bird breeding colonies on both sides of the Atlantic would have been like if they had not been subjected to continual human persecution:

> The rocks viewed from the sea, present a grand and very interesting spectacle, exhibiting masses of inclined, perpendicular, and projecting cliffs, smooth, largely fissured, or minutely intersected. Their whole face, along an extent of half a mile, was covered with birds, of which, notwithstanding their immense numbers, there were only four species: the guillemot, the auk [i.e. Razorbill], the puffin and the kittiwake. These birds inhabit the cliff, not promiscuously, but with a degree of regularity and distinction which strikes the beholder with wonder...Of the auks and guillemots, which lay only a single egg, placed on the bare rock,[6] one may often see on a shelf not more than three yards in length and as many feet in breadth, fifty or sixty individuals, jammed together like a solid mass, and each sitting upon its egg. Such masses are of frequent occurrence, the shelves being larger or smaller, but in general two or three or four are seen together, and sometimes an individual is observed sitting solitary, if one may say so, when it is surrounded by others at no greater distance than three or four feet...When a shot was fired, most of the birds in the neighbourhood left their stations and flew about, while some fell into the sea, and on emerging raised with their wings a continuous sheet of spray which extended several hundred yards from the rocks. After a succession of shots, almost the whole body seemed to be on [the] wing, presenting the appearance of a cloud, which occupied a quarter of a mile square, and through which one could scarcely distinguish the blue sky from the flakes of clouds. In their flight the birds did not cross much, but generally moved in the same direction, wheeling in a large circle, a disposition which probably arose from their number being so great that they could not conveniently fly at random. Their mingled screams were blended into one harsh mass of sound, in which the cry of individuals could not be distinguished. The noise and bustle reminded us of some great city, and the prodigious number could be compared to nothing but the shoals of some species of fish. Some were fishing on the smooth sea around the island, some flying from the rocks, some resting along the margin of the water, upon shelves or projecting crags, while by far the greater number were sitting upon their eggs. Such was the appearance of the place when the birds were not disturbed, and they were by no means very excitable, for unless after a shot, none stirred on our account, however close the boat came.

In England, where the pressure on the sea bird breeding colonies was greatest, there was some dispute as to the reason for the decline. In 1864, Colonel Henry Newman, who was in the habit of taking his son to Flamborough Head for the pleasure of shooting guillemots and the other sea birds referred to by Rennie, addressed the following plea to the editor of *The Zoologist*:

> Yorkshire is a sporting county and why do not the gentlemen unite and raise a small fund to prevent the collection of the eggs at unseasonable times of the year, that is, during April, May, and until the 24th of June? But for the great height of these cliffs the whole of the birds would be destroyed. At Filey it was stated that the quantities of eggs for all the species of birds mentioned, besides ducks, cormorants and gulls, are sold in the neighbourhood and town and the people say they are as good for the table as hens' eggs: a good many of them are sold as specimens. The birds might still be shot from the water, but that would be a trifling matter compared to the destruction of their nests and eggs; and if a stringent rule be made to enforce the preservation of these, no good sportsman or right-thinking man would trespass at improper seasons of the year.

John Wolley, as related in Chapter 11, had advocated that legislative steps be taken to safeguard the traditional rights of those, known locally in Yorkshire as 'climmers', who made a living from harvesting the eggs of sea birds — men whose activities he believed made no serious inroads into the breeding populations. Nothing could be as diametrically opposed to John Wolley's considered opinion as that expressed by Colonel Newman and it is difficult to avoid the conclusion that the latter's judgement was affected by an unhealthy snobbery. The riposte was delivered by Mr N.F. Dobrée of Hull, in the next issue of *The Zoologist*:

> As regards the taking of the eggs, the danger of obtaining them limits the climbers to a very small number — this year, I believe, only four.[7] These men, looking on it as a sort of monopoly, are as anxious for the preservation of the birds as any ornithologist can desire: they parcel out the line of cliffs amongst them at the opening of the season, each taking about a mile, and are as systematic in their mode of procedure as any Icelander in the collection of the eider down. They limit their descents over the same ground to two in number, not baring the cliffs entirely, but leaving rallying points for the birds to collect around again, and it is well-known that they will a third time renew the process of incubation without the number of eggs being seriously impaired. In the face of these facts, any action on the part of the landowners in preventing the taking of the eggs would unfortunately serve the cause only in an extremely partial degree; the high seas would continue open to all, and the boatmen below, who look upon the shooting parties as a large item of the summer's profits — and the birds as one of their vested interests — would remain as open to the allurement of a half-sovereign as before, although they one and all admit they are killing the goose for the sake of the golden eggs.

There exist some accurate figures for the numbers of sea birds which perished in just one year on that part of the coast of Yorkshire — figures which show that the scale of what can only be described as gratuitous killing was truly alarming. At the same time the latest fashion, of adorning hats with birds' feathers, had become 'the rage' in the English-speaking world. *The Guardian* newspaper, in its edition of 18 November 1868, included the following bleak paragraph:

Fig. 3 A Yorkshire 'climmer' with his outfit. (Photograph by T.H. Nelson)

> On a strip of coast eighteen miles long near Flamborough Head, 107,250 sea-birds were destroyed by 'pleasure parties' in four months; 12,000 by men who shoot them for their feathers to adorn women's hats and 79,500 young birds died of starvation in emptied nests. Commander Knocker, there stationed, who reported these facts, saw two boats loaded above the gunwales with dead birds, and one party of eight guns killed 1100 birds in a week.

It is well known that the cruelties of the feather trade did much to bring about legislation for the protection of birds on both sides of the Atlantic. It is, therefore, of interest to note that the destruction inflicted by the 'sporting' fraternity on this stretch of the English coast was *nine times* as severe as that caused by those who killed to supply the plume trade.

In the United States of America it was necessary to call on the great tradition of voluntary effort to pound on the doors of the state legislatures and so achieve, piecemeal, the protection of species whose breeding success was annually jeopardized by scores of feather hunters. Bitter and unscrupulous opposition from those with vested interests who felt their pockets threatened by increasing numbers of 'Audubonites', as those in favour of bird protection were disparagingly termed, was an ugly feature of the campaign. In Great Britain such opposition as there was to proposals for bird protection came from that quarter of the political spectrum which always resents any interference with a freedom which has hitherto been enjoyed 'without let or hindrance'. In the main, however, parliamentary acquiescence was secured with noticeable ease because the shocking figures published in the newspapers, more than anything else, effected a sea change in public opinion.

Happily, Great Britain is of a size whereby nothing of any importance which happens in one part of the country can truly be said to be so far away as to be 'no concern of mine' — a common problem in geographically larger countries. Hence, there is the need for only a minimum of local variation in matters of law. As a 'unitary' state, moreover, Britain was — and still is — a country where the single-issue pressure group can hope to achieve its legislative objectives in a relatively short space of time; by contrast, lobbying to get anything done at the federal level in the United States of America is much more arduous. In this we can observe a wonderful paradox at work in the body politic of that stately Leviathan: as early as 1817 the progressive spirit had overflowed in the Massachusetts State Legislature into a bill affording protection to all bird species, not just those deemed useful to man. This law, when passed, created such a furore in the ranks of sportsmen that it became subject to a series of countervailing amendments enabling numerous species to be classified as game. However, the real battle for bird protection in the United States only reached its height in the years after 1910 — a year in which *The New York Times* could describe a bill to choke off the trade in the aigrettes of the Great White Heron in that state as 'piffling legislation'. That year, it so happens, was the thirtieth anniversary of the Act passed in Great Britain which consolidated the provisions of the first two Acts for the preservation of breeding birds passed in 1869 and 1872 respectively. It is not necessary to argue about the relative merits of each political system, merely to note that, even if it is late off the mark, the British Parliament can act with commendable vigour.

As Britain was, at that time, a more deferential society than latterly, members of the public were apt to listen attentively to pronouncements of various 'august bodies'. The British Association for the Advancement of Science — jocularly referred to by Alfred Newton in his correspondence as the 'British Asses' — regularly held well-publicized conferences at the end of August. On 22 August 1868, the Revd F.O. Morris listened to a speech delivered by Newton (who by that time had been a Professor at Cambridge University for two years) at the Association's meeting at Norwich. Realizing the importance of what he had heard, Morris was moved to write immediately to *The Times* instead of waiting for the newspaper's correspondent to file his report. Newton's theme had been much the same as that of John Wolley nearly two decades previously, but in view of the slaughter that had gone on unabated in the intervening years, his tone, as will be related in the following chapter, was even more impassioned than that of his late friend. In his letter to the newspaper's editor Morris was able to supplement Newton's melancholy descriptions of the plight of sea bird colonies on the coasts of England with information of his own:

> I am sorry to say there is worse in the background. I can assure you that it is the truth that one of these unfortunate birds [a kittiwake] was picked up alive on the beach near Bridlington, some half dozen miles from Flamborough Head, with both its wings...torn off, and another was picked up in a field alive with its bill tied so that it could not feed.

Notes

1 The Act of 1845 was superseded in 1859 by *An Act for the Protection of the Breeding of Wildfowl and the Preservation of Game.* In it, egging and fowling on Funk Island were specifically banned with a ten-shilling fine imposed for every egg or bird taken. Provision was made, however, for the poor settler to take eggs and birds for immediate consumption by himself or his family.

2 John Milne, writing in 1875, in attempting to account for the disappearance of the bird whose bones he had collected, stated, '[I]t is by no means a necessity that there should be a large annual production for the preservation of a species. This is seen in the American Passenger Pigeon where a stock of millions of birds is kept up by parents bringing up seldom more than two young ones annually.' If anyone at that time had predicted that within forty years the Passenger Pigeon would have followed the Great Auk into extinction, Milne would have been incredulous — yet such, as is well known, proved to be the case.

3 A member of the field club, Dr Edward Charlton, had written what Tristram described as 'an able paper' on the species. Although it is in many ways commendable, the paper failed to recognize the economic factors at work in the decline of the species and remained ambivalent about whether it still existed. With disarming frankness this author concluded: 'It is possible that a few of these birds still survive on the islets of Newfoundland or Labrador; but, if not already extirpated, the Great Auk will, ere many years have elapsed, be numbered amongst the things that were.'

4 Spelt 'Fern'.

5 St Cuthbert was bishop of Lindisfarne 673–86.

6 The Razorbill generally lays its egg in a crevice.

7 The farmers who owned the land above the cliffs passed on the right of collecting the eggs to their workmen. In this way the practice was monopolized by a handful of families — principally the Londesboroughs, the Leppingtons, the Hodgsons, the Hartleys, and the Foxes — whose fitter members were lowered in leather harnesses designed to prevent them dangling upside down should they have the misfortune to knock themselves unconscious on the cliff face.

CHAPTER 16

An Act of Parliament

The paper[1] read by Professor Alfred Newton to the biology section of the British Association for the Advancement of Science on 21 August 1868 has traditionally been regarded as the immediate cause of the successful passage through Parliament of the first wild birds' protection bill — a law specifically proposed for the protection of sea birds at their breeding sites — which reached the statute book on 24 June 1869.

Significantly, Newton stressed in his opening remarks the important part that public attention to such matters could play: 'The most effectual protection to animals,' he stated, 'is that afforded by public opinion.' Surprisingly, perhaps, he gave by way of illustration the example of the fox, a species which was 'no longer considered merely a pestilent marauder of hen-houses', but was recognized as playing a useful role as a natural regulator, maintaining the balance of nature. The Act passed in Newfoundland in 1845 to protect breeding wildfowl had come into law, as had the proclamations which preceded it, solely because sea birds were regarded as useful as indicators of the depths of the waters off the coast. It was a brave man who would depart from these strictly practical considerations, yet it is clear that this is what Newton, in constructing a speech which would appeal at once to his immediate listeners and to a wider public, was minded to do. The possibility of future exterminations was evidently uppermost in his mind: 'Man,' he said, 'had no spite against the bustard[2] or the great copper fly'[3] — but both had been extirpated within living memory, the latter largely owing to the drainage of the fens. He added: 'Both, however, might possibly have been preserved by a little judicious care. At any rate if the progress of Civilization unconsciously demands some few victims, we should abstain from wilfully adding to their number.'

It is difficult to believe that Newton was not tempted to mention the Great Auk at this point, but to have done so would have been to digress to a subject about which the majority of his listeners, and certainly the wider public, would have been ill informed.

Having introduced the moral dimension of his first theme, Alfred Newton immediately reverted to practical considerations. This time, however, he intended to show that those species which were widely persecuted on account of having hooked beaks and claws, did, like the fox, have their part to play in the balance of nature:

> Mr Tristram contended at the last meeting of the Association [at Dundee in 1867] that birds of prey were the sanitary police of nature, and that if they had existed in their original strength, they would have stamped out the grouse disease just as the Orders in Council stamped out the cattle plague. The hawk by preference makes sickly birds its quarry…I believe the abundance of game has little to do with the scarcity of birds of prey, and can declare that in some foreign countries the existence of numerous birds of prey is a pledge of the plentifulness of game. Owls are the game-preservers best friend. His most serious foe is the rat and the owl consumes more rats and mice than any other description of food — and so it is with regard to polecats, stoats and weasels.

Intended for leisurely consideration by a wider audience, this passage nevertheless seems like an indirect compliment to Henry Baker Tristram for the part he had played behind the scenes in pressing for bird protection.

The most significant and radical section of Alfred Newton's speech is that concerning sea birds which were intended by the recently founded Close-time Committee to be the first beneficiaries of statutory protection. With considerable political acumen — appropriate in the son of a former Tory MP — Newton chose to 'create a stir' by focusing exclusively on the perceived moral and emotional aspects of the question:

Fig. 1 Portrait of Professor Alfred Newton (1829–1907) in his later years.

Now for sea-fowl — and here I must plead guilty to the charge (if it be a charge) of being open to a little bit of sentiment. At the present time I believe there is no class of animals so cruelly persecuted as the sea-fowl which throng to certain portions of our coast in the breeding season. At other times of the year they can take good care of themselves, as every gunner on the coast knows; but in the breeding season, in fulfilment of the high command to 'increase and multiply' they cast off their suspicions and wary habits and come to our shores. No-one that I have ever heard of has complained of them as injurious in any way. Some few, as the 'Scoulton Peewits' [Black-headed Gulls] settle far inland, and their usefulness as they follow the plough is everywhere recognised. But of the rest — I never heard the Willocks [Guillemots] or Kittiwakes of the Yorkshire coast accused of raising the price of herrings, sprats, and oysters! I think we may fairly assume that they are innocuous in every respect. But how do we treat them? Excursion trains run to convey the 'sportsmen' of London and Lancashire to the Isle of Wight and Flamborough Head where one of the amusements held out is the shooting of these harmless birds. But it is not merely the bird that is shot that perishes — difficult as it is to say where cruelty begins or ends — that alone would not be cruelty in my opinion. The bird that is shot is a parent — it has its young at home waiting for the food it is bringing far away from the Dogger Bank or the Chops of the Channel — we take advantage of its most sacred instincts to waylay it, and in depriving the parent of life, we doom the helpless offspring to the most miserable of deaths, that by hunger. If this is not cruelty, what is? Can men blaze away hour after hour at these wretched inoffensive birds and call it 'Sport' without being morally the worse for it? We thank God that we are not as Spaniards are, who gloat over the brutalities of a bullfight. Why, here in dozens of places around our coasts, we have annually an amount of agony inflicted on thousands of our fellow creatures, to which the torture of a dozen horses and bulls in a ring are as nothing. Surely I may be pardoned if I indulge in a bit of sentiment here? I began by deprecating over-coloured statements, or I might dwell on this ghastly picture much longer, but there is one painful feature which it is said has been lately superadded. The modern fashion of ladies wearing plumes in their hats is said to give an impetus to the slaughter. This rests on good authority. Mr [John] Cordeaux writes of the Kittiwake at Flamborough (*Zoologist* p. 1009):

'This graceful and trustful bird is threatened with speedy extinction at this famous breeding place; thousands have been shot in the last two years to supply the "plume trade". The London and provincial dealers *now* give one shilling per head for every White Gull forwarded; and the slaughter of these poor birds during the season (the *breeding* season remember) affords almost constant and profitable employment to three or four guns. One man, a recent arrival at Flamborough, boasted to me that he had in one year killed, with his own gun, four thousand of these gulls; and I was told that another of these sea-fowl shooters had an order from a London house for ten thousand.'

No wonder the kittiwakes are rapidly disappearing. There has this year been a marked diminution of the great breeding colony in the Speeton cliffs. Fair and

> innocent as the snowy plumes may appear in a lady's hat, I must tell the wearer the truth — 'She bears the murderer's brand on her forehead.'
>
> Now, that a stop should be put to this wanton and atrocious destruction of a species, aggravated as it is by circumstances of peculiar cruelty, I think none of my audience will deny. The only question is how it should be done. As I have said before, no doubt public opinion would be the most effectual check; but on the other hand, I fear lest by the time we can hope to influence public opinion to such a degree that laricide shall be regarded in the same light as vulpicide, there will be no more kittiwakes on our coast to protect. It seems to me, after due reflection, that legislative interference is absolutely required, for we can hope to influence Parliament in the matter sooner than we can that of the nation at large. And this brings me to the special object of this paper. In many countries as you are aware, there is a 'close-time' declared by the local authorities, during which the mere act of carrying a gun is an offence against the law. I need scarcely say that this 'close-time' extends over the breeding season. If the present state of things continues much longer, changes will occur with regard to our fauna for which we shall receive few thanks from our posterity.

Alfred Newton had measured his words well: the correspondent of *The Times* wrote that he had 'almost electrified his fair hearers' with his reference to the biblical story of Cain who murdered his brother Abel.[4] A vigorous intervention from the floor was made by Miss Lydia Becker of Manchester who protested against holding ladies responsible for the 'murder' of White Gulls:

> Much of the mischief in this world is done through ignorance: no woman will *willingly* wear the feather of a bird that has been destroyed in the act of feeding its young. Women should be instructed on these and other subjects and should be allowed to meet with the opposite sex on equal terms — not merely as listeners in on the discussion and in the acquisition of various branches of knowledge in which we are all interested — instead of meeting, as we do, with discouragement. If what I am suggesting is pursued, naturalists will soon find they have no reason to complain about the conduct of ladies.[5]

Miss Becker sat down to considerable applause.

Leave to introduce a Sea Birds' Preservation Bill into the House of Commons was granted in the following February. The parliamentary sponsor was the member for the East Riding of Yorkshire, Christopher Sykes. It was necessary to have agreement between the Close-time Committee and their sponsoring MP on what arguments would, on a free vote, gain most cross-party support. Traditionally, the preamble to any Bill outlines the reasons why it is considered necessary to have a new law enacted. It is significant, therefore, that the exact form that the preamble should take was left to be decided only after the principle of the Bill had been accepted by the House. Clearly it was felt expedient not to risk provoking controversy by stating the reasons for the Bill too baldly. These had been transformed, by the only lawyer on the Close-time Committee, Edmund Harting, into a carefully crafted preamble. Although it was, in the event, not used, it gives a clear insight into the way in which the minds of the Bill's architects were working:

> Whereas large numbers of sea-fowl are wantonly every year taken and destroyed on the coasts of England and Wales, without any regard to season, whereby the extinction of such sea-fowl is threatened: and whereas the said sea-fowl are of signal service to mariners on the said coasts in warning them of rocks, shoals and other dangers of the sea; and also to the fishermen on the said coasts in directing them to the most productive fishing grounds; and the said sea-fowl are also of use for sanitary purposes in removing various kinds of offal from the harbours and shores of seaport towns; and further that the said sea-fowl contribute much to the beauty of the coast scenery: and it is therefore expedient to protect such sea-fowl during the breeding season, and to prevent their eggs from being taken or destroyed: Be it enacted [etc. etc.]

There are traces here of the controversies which must have been argued over in the Close-time Committee and which were duly taken up by certain members of the House of Commons. If the Scottish breeding grounds of the Guillemot and other Auks were more populous and less threatened than those in England, there was a strong argument for restricting the operation of the Act to England and Wales only. Furthermore, to have singled out the day-tripper from the big towns and to have prevented his so-called sport while still permitting the eggers to pursue their equally time-honoured practices unimpeded, would have been unwise. In terms of the damage inflicted, there was a strong case for doing so, but practical politics seems to have demanded that both practices should be prohibited if the Act was not to prove socially divisive. However, it is difficult to imagine either Alfred Newton or the Revd H. B. Tristram — oologists both — approving such a clause except with extreme reluctance. In the event, as will be related, a successful rearguard action was fought in Parliament to safeguard the privileges of the Yorkshire 'climmers'.

The Act passed by the Newfoundland Assembly in 1845, as we have seen, had concentrated on the usefulness of birds — it certainly would have been most unexpected if the draftsmen in St John's had waxed lyrical about the contribution of sea birds to the beauty of the coastal scenery! Perhaps Mr Sykes advised that a similar approach was best calculated to be successful with the hard-bitten men of the world who made up the bulk of the House of Commons in 1869. Rising to his feet on 26 February of that year, Christopher Sykes addressed his fellow MPs as follows:

> It is with much diffidence that I presume to bring before the notice of the House the measure which ought to give some legislative protection to those sea-birds which still remain off our English coasts. The Bill is not only framed in

accordance with the strongly expressed feeling of almost every class of my constituency but from the numerous letters I have received from all parts of England, evincing the warmest sympathy with its objects, I am led to regard it as one of almost national interest. The sea-birds of England are rapidly disappearing from our coasts. This fact was established at the last annual meeting of the British Association. From Northumberland, Durham, Yorkshire, Norfolk, Cornwall and Pembrokeshire the same cry arises. My Bill aims at protecting those sea-birds in the breeding season; and there is a very important precedent for it in the statute passed in the 25th year of the reign of Henry VIII protecting sea-birds[6] at this very season. The grounds on which I bring the measure forward are no mere sentimental or humanitarian grounds, though these are strong enough. I do so in the interest of three very important classes of my constituents, for very important they must be to every Member of a seaboard county. I mean the farmers, the merchant seamen and the deep-sea fishers.

A few years ago the farmers of the East Riding of Yorkshire — not merely those residing in the immediate vicinity of the coast but as far as twenty miles inland — were accustomed to see flocks of sea-birds following in the heels of the plough-boy and from the newly turned up earth picking up worms and grubs. But I have here a letter from an influential farmer living in the parish of Filey, within a mile of the coast, stating that last summer he did not see a single bird on his farm. I appeal to the House also in the interest of the merchant sailors, for in foggy weather those birds, by their crying, afford warning of the proximity of a rocky shore when neither a beacon light be seen nor a signal gun heard. I have here a paper proving that with the decrease of these birds, the number of vessels which have gone ashore at Flamborough Head has steadily increased. For the services they render to the mariner, these birds have earned for themselves the name of 'Flamborough Pilots'. I appeal to the House, likewise, in the interest of the deep-sea fishers because, by hovering over the shoals of fish, these birds point out the places where the fisherman should cast his net. On this ground alone the legislature of the Isle of Man has lately passed an Act imposing a penalty of Five Pounds on every man who wilfully kills or destroys a seagull. Lastly I make my appeal even in the interest of those thoughtless pleasure-seekers themselves who flock to the coasts in the summer months, chiefly from the populous towns of the West Riding of Yorkshire and Lancashire. Those persons will have themselves to blame, if in a few years, those rocks which once I remember as teeming with wild-fowl, have become a silent wilderness.

I hope the importance of the Bill which I move for leave to be introduced, will not be over-looked through the insignificance of its advocate.

On the 5th of March, 'disclaiming all acquaintance with the natural history point of view', Mr Sykes moved a motion for the second reading of the Bill. In this he was supported by another member of the House, O. Stanley, who focused on the 'wanton cruelty of destroying sea-birds' as well as on the prevention of many shipwrecks. In support of his argument he quoted from a letter from the Deputy Master of the Trinity Lighthouse Board, Mr F. Arrow, who had expressed the wish that the Act could be extended to prevent the taking of eggs also. Mr Stanley concluded: 'With regard to the latter point, I understand the honourable promoter of the Bill is willing to include eggs within its provisions.' Accordingly, Section 4, prohibiting the selling of sea birds' eggs, was inserted into the Bill and accepted by the House of Commons — though not, as will be related hereafter, by the House of Lords.

The only note of controversy at the second reading in the Lower House was an attempt, unsuccessful in the event, by a number of Scottish members who hoped to exclude their part of the United Kingdom from the provisions of the Act. They considered over-exploitation of sea birds to be a peculiarly English problem — they conveniently forgot about the Gannets of the Bass Rock — and not, in their view, one which warranted imposing penalties on the poorer islanders who sought to supplement their income or diet by carrying out raids on the immense gatherings of sea birds there. This potentially divisive issue was also debated in the House of Lords so that, ultimately, the St Kildans were not put in the invidious position of feeling obliged to flout the law if they wished to continue their ancestral way of life.

Realizing that sea birds went by a bewildering variety of local names, the drafters of the Bill attempted to prevent anyone from escaping conviction on account of his own (or the magistrate's) ignorance of a species' most widely used vernacular name.[7] The main thrust of the Act was to be found in Section 2 which imposed a maximum fine of one pound for every sea bird 'killed, wounded or taken' between the 1st of April and the 1st of August in any year. There were also penalties for being found in possession of such birds and for using any 'boat, gun, net, or other engine or instrument' for the purpose of killing, wounding or taking any sea bird.

By ensuring that the fines imposed were proportionate to the number of birds killed, this United Kingdom law was consistent with an amendment dating to 1859 of the first Newfoundland legislation. The Act of 1845, by imposing liability at a twenty-pound fine for each and every bird killed, had been looked upon by the islanders with undisguised hostility so that it became necessary — or expedient — to reduce the scale of the fines, making them proportionate to the number of birds killed, and to permit a man to take sea birds for the immediate consumption of his family. Perhaps nothing illustrates the apparent disregard with which the earlier legislation was viewed than that the amended law of 1859 made it specifically illegal to go egging or fowling on Funk Island. Any law of this kind, however, requires the co-operation of the public and we may imagine that the Newfoundland fisherman frequently turned a 'blind eye' to the actions of his neighbour. Aware of the imperfect application of the law in Britain's oldest colony, those who framed the Act for the Preservation of Sea Birds included, through Section 5, the following provision:

> One moiety [half] of every penalty and forfeiture under this Act shall go and be paid to the person who shall inform and prosecute for the same, and the other moiety shall, in England, be paid to some one of the overseers of the poor…of the parish, township, or place in which the offence shall have been committed.

The impact of such a clause was, as mentioned in Chapter 10, to have a dramatic effect for the better on sea bird populations.

The Duke of Northumberland sponsored the Bill in the House of Lords — a chamber which at that time, it should be remembered, was equal in power and authority to the House of Commons. Here, there was some controversy regarding a number of points already debated in the House of Commons; the Duke of Richmond, evidently disgusted by the behaviour of the shooting parties on the coast of his native Yorkshire, attempted to substitute the words 'shoot or attempt to shoot' for the more inclusive phrase which had been agreed by the Commons:

> For if the inhabitants of the coast were allowed to capture birds by snares and other ingenious but difficult contrivances, there would be no fear of their numbers being sensibly diminished. In the islands off the coasts of Ireland and Scotland, the inhabitants maintain themselves for a considerable portion of the year on the eggs and food [*sic*] of these birds, and at St Kilda this has been the case for upwards of two hundred years without having had the effect of diminishing their numbers.

The Duke of Northumberland, mindful of the arguments advanced by the Scottish members in the House of Commons, already had a plan, which in due course became Section 8, to exempt St Kilda from the provisions of the Act. St Kilda was unique for a number of reasons, not least its distance from the mainland, which rendered enforcement of the Act a practical impossibility. In the end, making St Kilda a special case[8] was justified on the grounds that it was doubtful, in the words of the Duke of Argyll, whether there was another island community in which the inhabitants 'practically lived on' sea birds.

On the question of whether the Duke of Richmond's proposed amendment should be accepted, there was a vigorous intervention by the Archbishop of York: 'The Bill might as well have been rejected if so serious an alteration is to be made in it. The principle of the Bill is to prevent the birds being killed in any way during the breeding season.'

In a noticeably contradictory fashion the Duke of Richmond defended the plume hunters from being regarded in the same light as the (other) thoughtless pleasure seekers, pointing out that fashions do change. However, he neglected to address the critical question of whether local extinction might not occur first. It is as well that he withdrew his amendment, finding the mood of the House against it, for the habit of 'the fair sex' of adorning their hats with feathers was to persist until after the nineteenth century was out.

Turning their attention to the proposed Section 4 which imposed penalties for taking eggs other than for food, the Duke of Richmond, with paternalistic regard for the poorer inhabitants of his county, was once again to the fore:

> It would cause much suspicion and annoyance to the hundreds of poor people who take eggs for food and who might be repeatedly brought before magistrates to state that eggs in their possession were taken for that purpose. In ninety-nine cases out of every hundred, eggs are so taken, and in the hundredth case a man would naturally assert that he intended to eat the eggs.

Introducing a touch of levity into the proceedings he asked: 'How would the magistrate be satisfied of this intention unless the man ate them then and there in his presence? Moreover the clause would impose a penalty on a person who had received the eggs from a friend, while that friend, the real offender, would escape.'

This line of reasoning was taken up by Lord Houghton who, evidently speaking on behalf of 'the good and the great', specifically expressed an interest in the natural history aspect of the question:

> Among scientific men are some rejoicing in the peculiar name of oologists to whom the collection and classification of eggs is a subject of great interest. Under the clause a learned professor[9] of one of our great universities might be taken up by a policeman for having a sea-bird's egg in his possession for scientific purposes. The number of eggs taken for purposes other than those of food, is so small that it is hardly worthwhile to legislate on this point.

Supporting Lord Houghton, the Bishop of Oxford attempted to reintroduce a note of jocularity: 'The clause might operate harshly on the owners of collections of eggs. An ornithologist might have had a valuable egg in his collection for fifty years. Now, it would be very hard to insist on his showing that it was intended for food by eating it!'

Leaving mirth aside, the Duke of Northumberland defended the clause on the grounds that it had stood the test of considerable debate in the House of Commons. However, the Lords divided on the question and the motion to remove Section 4 was passed by 54 votes to 40 — a small but significant victory for the Duke of Richmond and the poorer inhabitants of the coast of Yorkshire.

The amendments put forward by the House of Lords were ratified by the House of Commons on the 2nd of June and the Act received the Royal Assent three weeks later. The preamble adopted was rather a bland affair when compared to the original draft, but it served its purpose: 'Whereas the sea-birds in the United Kingdom have of late years greatly decreased in number; it is expedient therefore to provide for their protection during the breeding season.'

Masterly as the Act for the Preservation of Sea Birds of 1869 undoubtedly was, it was not, alas, flawless. Although most of the Auk species have departed their

Fig. 2 Fledgling Kittiwakes at their nest on Lundy. (Photograph by Alan Richardson)

breeding grounds by August, a number of sea birds including puffins and kittiwakes, may be found still feeding their young in the middle of that month. Perhaps the sponsors of the Act thought they would have been 'pushing their luck' to have attempted to establish a breeding season of more than four months' duration. The eggs of the kittiwake 'are seldom laid until the last week in May, and sometimes not until the first week in June,' wrote Howard Saunders, the editor of the third and fourth volumes of the revised edition of William Yarrell's *History of British Birds* which appeared in the early 1880s. Consequently:

> ...many of the young are still in the nest, or barely fliers, when the Sea Birds' Protection Act expires on the 1st of August. Some years ago, when the plumes of birds were much worn in ladies' hats, a fashion which any season may see revived, the barred wing of the young Kittiwake[10] was in great demand for this purpose and vast numbers were slaughtered at their breeding haunts. At Clovelly, opposite Lundy Island, there was a regular staff for preparing the plumes, and fishing smacks with extra boats and crews used to commence their work of destruction at Lundy Island by daybreak on the 1st of August, continuing this proceeding for upwards of a fortnight. In many cases the wings were torn off the wounded birds

> before they were dead, the mangled victims being tossed back into the water; and the Editor has seen hundreds of young birds dead or dying of starvation in the nests, through want of their parents' care, for in the heat of the fusillade no distinction was made between old and young. On one day seven hundred birds were sent back to Clovelly, on another five hundred, and so on; and, allowing for the starved nestlings, it is well within the mark to say that at least nine thousand of these inoffensive birds were destroyed during the fortnight.

Evidently the passage of an Act of Parliament had done nothing to effect a fundamental change in the disposition of those west country fishermen towards the sea birds with which they were so familiar — but these were the men whose ancestors, we should recall, had plied their trade so assiduously off Newfoundland.

Notes

1 *The Zoological Aspect of the Game Laws*

2 The Great Bustard — the last indigenous British birds were recorded in 1838.

3 The English subspecies of the Large Copper *Lycaena dispar dispar*. In common with that of the Great Auk, the responsibility for the ultimate extirpation of this beautiful butterfly lay with collectors. It was last recorded in *c.*1850 at Bottisham, Cambridgeshire.

4 See Genesis 4: 15.

5 In the United States it was the wives of prominent ornithologists who commenced the agitation against the wearing of bird feathers. However, a considerable number of women persisted in wearing various plumes even when, how, and at what season of the year they were collected, had become common knowledge.

6 The statute was in fact intended to protect the wildfowl which, customarily, were served at royal banquets.

7 Sea birds were deemed to include 'every species of Diver, Grebe, Shearwater, Petrel — including the Fulmar — Skua, Gull, Tern, and Auk'. Specifically mentioned as well were the Cornish Chough, the Gannet, and the Oystercatcher and a number of sea duck including the Eider, the Merganser, and the Smew. Other sea duck and also a few sea birds properly so-called (notably the Cormorant and Shag) were omitted from the definitive class in Section 1.

8 In its final form, the Act (through Section 9) included provision for other parts of the United Kingdom to be exempt from its operation from time to

time, by Orders in Council, if this should be thought desirable for any reason.

9 What conversations may have taken place between Houghton and Alfred Newton prior to the reading of the Bill in the House of Lords must remain a matter of conjecture.

10 The so-called 'Tarrock' plumage.

Epilogue

In spite of the adoption in many countries of measures to protect rare and endangered wildlife, the list of species that have become extinct has continued to grow. The populations of others have dwindled with the encroachment of modern methods of farming, forestry, and fishing. Early in the twentieth century Walter Rothschild, reflecting on his unrivalled collection of rare and extinct birds housed at Tring, north of London, observed: 'The melancholy fact remains...that man and his satellites, cats, rats, dogs and pigs are the worst, and in fact the only important, agents of destruction of native avifaunas wherever they go.'

How far human ignorance and stupidity are to blame — rather than wantonness, cruelty, and greed — depends on the circumstances of each case. However, we can be thankful that, when seen in relation to population growth and advances in technology, the rate of known extinctions in the industrialized West has not increased exponentially over the years. Undoubtedly, some species, which but for protective measures would have followed the Great Auk into extinction, have been saved through the agency of concerned individuals and societies. The establishment of private and government sponsored conservation bodies, international conventions, the very high standard of wildlife documentaries, as well as the impact of two calamitous wars, have all played their part in transforming public attitudes in developed countries towards living creatures. For centuries previously the contact of Europeans with newly colonized lands was marked by an aggressiveness of which we are now ashamed.

A generation before the publication of Darwin's *Origin of Species*, educated people were shocked by the indisputable evidence adduced by the French anatomist, Baron Cuvier, that in the remote past there had lived large animals that no longer existed. Many were demoralized by the realization that their understanding of the created order was in need of drastic revision if the apparent prevalence of the process of extinction in nature was to be accounted for. After 1859, when Darwin's book 'came into the theological world like a plough into an anthill', the idea of extinction came to be regarded as part of the natural order itself. As Darwin himself remarked, 'No fact in the long history of the world is so startling as the wide and repeated extermination of its inhabitants.'

Nothing had, as it were, a lien on eternity and every species of animal, bird, and plant would in time either die out or evolve into some more advanced life form. Few today, however, will deny that even one preventable extinction is as the loss of a precious gem once set in a diadem of matchless beauty.

The extent to which prevailing opinion has fundamentally changed in recent decades may be gauged by a passage from *In Praise of Birds* (1925) by the English churchman and peace activist, Canon C.E. Raven:

> It is obvious that with a population like ours we cannot regret the efforts that have changed the Cambridgeshire fens into rich ploughland, or the salt marshes of North Norfolk into pasturage, or the sand dunes of Lancashire into golf links, or the Broads into a holidaymakers' paradise. That many and in particular the larger birds should be driven elsewhere is inevitable. The price has been paid in the interests of the majority of our people. It is misplaced sentimentalism to lament it.

The problem we face today is that the 'elsewhere' to which Canon Raven suggested larger and more wary species could go to seek refuge are hardly to be found. Man's endowment with an ethical sensibility and his capacity to choose whether he will continue in time-honoured self-centredness or reappraise his role in the natural order was raised by Fairfield Osborn in his epoch-making book *Our Plundered Planet* (1948):

> The animals of the earth are dependent on man's sufferance now, and of late years conditions have improved for them. There is a growing public consciousness that whether or not we need them for utility's sake, we must protect them. National Parks and refuges have been set aside for them in the States and in many countries. But civilisation and the rising needs of increasing numbers of people are pressing hard upon the last wildernesses. Man must live, but one wonders sometimes, in far away moments, whether there is not a primal form of ethics involved. Should not man perhaps, even for his own peace of mind, think of himself not as the consumer alone but as the protector? Like companions of an earlier life, if we forget them they are gone forever. It is man's earth now. One wonders what obligations may accompany this infinite possession.

Where the environment continues to be degraded either by commercial interests or by haphazard development, there are steps which, perhaps only with difficulty, national governments can and must take. The real challenge, however, in certain quarters remains a public opinion at once ignorant of and indifferent to the environment in which they live. Lack of will to ameliorate this state of affairs means that the 'last wildernesses' are increasingly littered with non-biodegradable waste, their delicate ecosystems scarred by four-wheel drive vehicles or, it may be, reduced to ashes by the indiscriminate use of 'slash and burn' methods of agriculture. Governments must accept the obligation to create a better informed public at home and to place environmental matters on the high table of international discussion. In order that policy is not to be seen as in any way anti-human, a careful distinction must be made between what we do and how we do it. Habits which were unexceptionable when populations were smaller and technology less advanced are now devastating. The lessons must be learned by the directors of logging companies as much as by the humble farmer with a family to feed. Newspaper proprietors have a positive role to play in this process too, although there exists a danger that in sensationalizing the issues their editors may risk alienating many readers.

The history of the loss of the Great Auk sheds light on tragedies which might still be averted. Disaster will surely result wherever Government fails to regulate the abuse of that unfettered economic activity beloved of political libertarians. Let no one say that legal regulation is an unwarrantable infringement of human rights and liberties when such legislation is concerned solely to stave off the long-term consequences of short-term human planning. Legislation passed without the political will to enforce it, however, makes a mockery of both Government and governed. In 1786, it will be recalled, further destruction of sea birds on their breeding grounds off Newfoundland was prohibited. Even if the great majority of the population respected the proclamation, such a legal instrument could be no guarantee of a species' survival if even one man in a hundred was prepared to ignore it. Yet the presence of a single naval vessel for six weeks of the year in the waters adjacent to the Great Auk's ancestral breeding site would have saved that species from the roving crews of egg and feather plunderers. We can only conclude that the extirpation of the Great Auk was inevitable if we allow that the worst of men will always find a way to accomplish their worst.

The Great Auk's vulnerability to extinction on account of both its flightlessness and its dependence on a tiny number of islands during the breeding season means that its plight is not typical of that of the majority of those species whose continued existence is still threatened. The deleterious effects of agricultural spraying, the destruction of breeding habitat, the consequences of oil pollution at sea — some of it deliberately caused — receive their measure of coverage in the news media. To these must be added the disruption of migratory strategies of many species as a consequence of climatic change and, in regard to sea birds, the loss of food supply resulting from the depletion of fish stocks. While the establishment of national reserves can safeguard the survival of certain localized species, it is not a solution to the many dangers now threatening a range of species once considered not to be in need of protection.

Early in the twentieth century the Hungarian ornithologist, Otto Herman, characterized the collective impact of migratory birds, in particular, as an immense *economic* force acting on the natural order:

> Nature herself, if in tact, does not recognise either useful or noxious birds but regulates the number of individual birds in accordance with the order and condition of their life, this regulation being, to use a modern term, automatic:

> on the other hand, where the ordinary conditions of nature change, the proportion of bird species changes in accordance with the variation in the supply of nourishment.
>
> If we survey all these points, we may get a true if somewhat hazy picture of the significance of birds...which neither Nature nor man can dispense with with impunity, the loss of which cannot be supplied by any manner of artificial procedure.

Like canaries once taken down coal mines to give warning of imminent danger from gas, the species that share our environment with us are a measure of its relative health.

We lament that in Newfoundland in the days of sail the King's Writ hardly ran to one small island crowded with sea birds. The processes which then led to extinction have changed little. They must be halted not only because humans are impoverished and demeaned thereby. Multicausal as these processes appear to be, they share one feature in common — the continuing human preference for short-term advantage over long-term gain. Increasingly this will be seen as the dilemma of democracy as party politicians face the ever-present temptation to appeal to the appetites of the electorate. Ultimately there can be but one solution to this dilemma for the process of extinction is one which shows no partiality in the selection of its victims.

The comparative anatomist F.A. Lucas, returning from his visit to Funk Island in 1887 for the purpose of collecting Great Auk bones, drew a clear distinction between the way things ought to be and the way they all too often are when the survey ship *Grampus* put into the Magdalen Islands in the Gulf of St Lawrence. The natural fecundity of these islands in the sixteenth century had been noted by the venturer Charles Leigh and published in 1600 by Richard Hakluyt:

> [T]hey do by nature yield exceeding plenty of wood, great store of wild corn like barley, strawberries, gooseberries, mulberries, white roses, and store of wild peason. Also, about the said islands, the sea yieldeth great abundance of fish of divers sorts. And the said islands also seem to proffer, through the labour of man, plenty of all kind of our grain, of roots, of hempe, and other necessary commodities.

The fir trees, Lucas remarked, had given way to scrubby spruce whilst:

> ...the great cods have become few and far between, the herring industry is completely unimportant, and a few short years have sufficed to seriously reduce the lobster industry. All this means want and distress for the population of these islands which, never too well off at best, has several times been saved from starvation by government aid, and once during the last 25 years forced to eat their very dogs.

In the decades since Fairfield Osborn wrote *Our Plundered Planet* the human race has come to realize its capacity to precipitate catastrophe on a previously unimagined scale. Let it not be said of us that we failed to adapt to meet the demands that this new reality has imposed on us.

Further reading

Bengston, S-A. 1984. Breeding ecology and extinction of the Great Auk (*Pinguinus impennis*): anecdotal evidence and conjectures. *Auk* **101**:1–12.

Burness, G.P. and Montevecchi, W.A. 1992. Oceanographic related variation in the bone sizes of extinct Great Auks. *Polar Biology* **11**:545–51.

Fisher, J. and Lockley, R.M. 1954. *Seabirds.* London.

Fuller, E. 1998. *The Great Auk.* Published by the author, Southborough, Kent.

Garðarsson, A. 1984. Fuglabjörg Suðurkjalkans. *Árbok Ferðafelags Íslands* pp. 126–60.

Gaston, A.J. and Jones, I.L. 1998. *The Auks* Pt. 1, pp. 3–117. Oxford, New York, Tokyo.

Kaufmann, L. and Mallory, K. 1993. *The Last Extinction.* Cambridge, Mass., and London.

Livezey, B.C. 1988. Morphometrics of flightlessness in the Alcidae. *Auk* **105**:681–98.

Lloyd, C., Tasker, M.L., and Partridge, K. 1991. *The Status of Seabirds in Britain and Ireland.* London.

Montevecchi, W.A. and Kirk, D. 1996. Great Auk. *The Birds of North America* (ed. A. Poole and F. Gill). Washington D.C.

Naumann, J.F. 1903. Der Riesenalk. *Naturgeschichte der Vögel Mitteleuropas* Bd. 12. Gera–Untermhaus.

Nettleship, D.N., Sanger, G.A., and Springer, P.F. (ed.) 1984. *Marine Birds, their Feeding, Ecology and Commercial Fisheries Relationships.* Ottawa.

Nettleship, D.N., Burger, J., and Gochfield, M. 1994. *Seabirds on Islands: Threats, Case Studies and Action Plans.* Cambridge, Birdlife International.

Olson, S.L. 1985. The fossil record of birds. *Avian Biology* **8**:79–238. New York.

Petersen, Æ. 1995. Brot úr sögu geirfuglens. *Náttúrufraeðingurinn* **65** (1–2):53–66.

Storer, R.W. 1960. Evolution in diving birds. *Proc. Int. Orn. Congress* 12:694–707.

Appendix 1

Description

The following description of a Great Auk, based on the specimen acquired in London by J.J. Audubon, was written by Professor William Macgillivray and was published in volume iv of Audubon's *Ornithological Biography* (1838) — a less expensive, separately issued text of *The Birds of America.* It is substantially the same text that appears in volume v of Macgillivray's own *A History of British Birds* (1852).

Adult in summer. –

The *body* is of a full and compact form; the *neck* short and thick; the *head* large, oblong, anteriorly narrowed.

Bill longer than the head, stout, very high, extremely compressed. *Upper mandible* with the dorsal line straight at first, then declinato-decurvate to the end, the ridge being narrow, broader at the base; the sides nearly flat, with a basal marginal ridge, succeeded by a deep narrow groove, then a large flat space, followed by eight oblique curved ridges and grooves, the edges sharp and direct towards the end, the tip decurved and rather obtuse. *Lower mandible* with the angle[1] long, the sides for half their length extremely narrow and linear, beyond the angle broad, that part being high and compressed; the dorsal line at first convex, then ascending and concave to the end; the sides flat with twelve transverse grooves, fainter than those of the upper mandible, the edges sharp, the tip deflected.

Nostrils basal, linear. *Eyes* rather small. *Apertures of ears* very small. *Feet* placed far back, short; *tarsus* short, compressed, anteriorly scutellate, scaly on the sides. Hind toe wanting; outer toe nearly as long as the third or middle, inner toe much shorter; all with numerous scutella and several rows of angular scales, and connected by narrow reticulated membranes; the inner and outer toes connected at the base, the middle toe only for a quarter of an inch. Claws rather small, narrow, arched, convex above, obtuse.

Plumage dense, blended, very short, on the head and neck short and velvety. *Wings* extremely small, but of the same form and structure as in the Razorbill and Guillemots; the *primaries* narrow and tapering to an acute point, the first longest, the rest rapidly graduated, their coverts long; *secondaries* short and broad, scarcely longer than their coverts. *Tail* short, pointed, of fourteen feathers.

> *Bill* black with the grooves white. *Feet and claws* black. The *head, throat, sides,* and the *hind part* of the *neck,* and all the *upper parts,* black; the *throat and sides of the neck* tinged with chocolate brown; the *wings* with greyish brown; the *head, hind neck,* and *back* glossed with olive green. A large oblong *patch before each eye,* the *tips of the secondary quills,* and all the *lower parts,* white.
>
> *Length* to the end of tail 25 inches[2] [63.5 cm]; wing from flexure 7 [17.8 cm]; tail 3 [7.6 cm]; bill along the ridge $3\frac{5}{12}$ [8.6 cm]; along the edge of lower mandible $4\frac{1}{2}$ [11.43 cm]; its depth at the angle $1\frac{7}{12}$ [4.02 cm]; tarsus 2 [5.08 cm]; middle toe $2\frac{8}{12}$ [6.77 cm], its claw $\frac{6}{12}$ [1.27 cm]; outer toe $2\frac{8}{12}$ [6.77 cm], its claw $\frac{4}{12}$ [0.85 cm]; inner toe $2\frac{1}{12}$ [5.29 cm], its claw $\frac{5}{12}$ [1.06 cm].

The following succinct description of the appearance of the Great Auk moulting to winter plumage was given by Dr Burkitt. The specimen (see Chapter 13) is preserved at Trinity College, Dublin.

> The head, back, wings, legs, and feet are sooty black; between the bill and the eye on each side of the head there is a large patch of white, mottled with blackish feathers; the neck is white, slightly mottled with black; the front of the body white; the lesser quills [i.e. secondaries] tipped with white.

The winter plumage of the Great Auk is imperfectly known. A specimen at the Zoological Museum at Copenhagen University, reputed to have come from Greenland, also has the large white spots in front of the eyes mottled with black. While this may represent the typical winter plumage, another specimen, preserved in Prague, has no visible white patch before the eye at all. It is possible, therefore, that the Copenhagen specimen was killed before it had acquired its full non-breeding plumage. As both of these specimens have the usual complement of grooves on the bill, it is unlikely that either of them is an immature bird.

Notes

1 i.e. the line of the gape.

2 This may be contrasted with Fleming's measurement of 3 feet for the bird he described in *A History of British Animals* (1828), the difference being accounted for by reference to the neck which could be extended or retracted at will. We might well expect a mounted bird, with its skeleton removed, to look considerably shorter — in other words, *very much as it would have appeared in life* when out of the water. Measurements from 32 to 36 inches are from tip of bill to tips of feet when the bird is at its most attenuated: this is useful for comparison with other, similarly sized, species but gives a misleading impression of how big the bird would actually have appeared. An incubating female Great Auk would have stood about 22 inches (approximately 58 cm) high; the largest males perhaps three inches more.

Appendix 2

Alca, *Plautus* or *Pinguinus*?

During the middle years of the nineteenth century, when European ornithologists were concerned to establish the range of the Great Auk and, more especially, whether or not it was still extant, the species was designated by the overwhelming majority of writers as *Alca impennis*— the flightless (literally the 'wingless') Auk. This was the name coined by the Swedish naturalist Linnaeus (*Fauna Suecica,* 1746) and, importantly, published in the tenth edition of his monumental *Systema Natura* (1758). This edition, in spite of revisions and changes to the names of certain species published in subsequent editions of his work, is taken as the starting point by taxonomists endeavouring to achieve international uniformity in the citation of scientific names.

The other closely related Alca described by Linnaeus was the Razorbill *Alca torda,* both species forming one genus ('tribe') of birds. However, Linnaeus' terminology has never been regarded as fixed for all time if it can be shown that emendation is warranted by advances, or simply changes, in scientific thought concerning what constitutes grounds for 'splitting' races to form new species, or species to form separate genera. Linnaeus' name in brackets after the Latin name for a species indicates that some such change in terminology has taken place. In this way *Alca impennis* Linnaeus is usually referred to today as *Pinguinus impennis* (Linnaeus).

Ever since the first half of the nineteenth century there has been a consensus, though by no means universal agreement, that on account of perceived differences in internal structure, the Great Auk and the Razorbill should form different genera. Change, of course, is always contentious and many will sympathize with Alfred Newton who, we are told, 'studied to avoid' questions of nomenclature for that very reason.

Chiefly, though not solely, because established usage is, in its own right, a strong argument for not effecting any change to the designation of recognized species, the name *Alca impennis* retained its champions long after emendation of the generic name was first proposed. In the late nineteenth century Elliot

Coues suggested, in *A Monograph of the Alcidae* (*Proc. Acad. Nat. Sci. Philadelphia*, 1868), that it was the Razorbill whose generic name should be changed (to *Utumania* Leach 1818) in order to preserve *Alca* for the Great Auk.

In establishing the 'type' species of any given genus, taxonomists identify which species within it was first described by Linnaeus. Another way to approach the question, equally conclusive where appropriate, is to examine the precise signification of the generic and specific names taken together: if both are found to be synonyms one for the other, taxonomists have an unanswerable case for deciding the type species. The Razorbill is a good illustration of the operation of both of these principles, for the specific name *torda* is derived from *tordmule* — the vernacular name for the species in Sweden. Accordingly Coues' arguments have never been accepted.

In the event of a proposal to dispense with a Linnaean name we confront another principle governing zoological nomenclature: the first generic name to be given by a subsequent taxonomist after the 1758 edition of Linnaeus' *Systema Natura* must, except under extraordinary circumstances, be used. In his monograph devoted to the species, Japetus Steenstrup addressed the question of which generic name was most appropriate for the Great Auk. In 1771, in his *Zoologiae Fundamenta*, Brünnich had proposed the name *Plautus* because the superficially similar birds of the Southern Hemisphere had been compounded under the generic name of Penguin — birds to which Brünnich himself had given the names *Pinguinus* and *Spheniscus*.

Ten years earlier the taxonomic record had been in an even greater state of fluidity. Linnaeus had proposed two genera, *Diomedea* and *Phaeton*, for the Southern Hemisphere birds, while Brisson (*Ornithologia* 1760, vol. 6) had devised a family in two genera — the Manchots (*Sphenisci*) and the Gorfue (*Cataractis*). These genera were distinguished by the shape of the apex of the lower mandible. Steenstrup, while inferring that Brisson's two genera and those of Brünnich were identical, and that Brünnich had in fact accepted Brisson's criteria, pointed out that the generic names advocated by both men, because they did not agree with the genera established by Linnaeus, could not have systematic priority. Moreover, Brünnich, in his *Ornithologia borealis*, had already employed the name *Cataracta* for a different genus.

In the first decades of the nineteenth century further attempts had been made to give a separate place to *Alca impennis*. Leach (as already mentioned) and G.R. Gray retained the name *Alca* — the latter only until a more suitable designation could be found. In 1856, Charles Lucien Bonaparte put forward the name *Pinguinus* (a name first used for the Great Auk in 1791 by Bonnaterre). Although Steenstrup had sympathy for Bonaparte's designation he was strongly of the opinion that to accept it would be to ignore the settled rules governing how generic names should be determined.

Although the name *Plautus* (in the alternative spelling *Plotus*) had been proposed by Linnaeus for the *Anhinga* instead of that of *Ptynx* (Möhring *Genera avium* 1752), Steenstrup argued that, in this instance, Linnaeus should not be followed and that the names *Plautus* and *Pinguinus* should be used in the sense that Brünnich employed them. Accordingly, in Steenstrup's opinion, as the representative of a monotypic genus, the Great Auk should be termed *Plautus impennis,* (Lin.). Steenstrup added that if this did not prove acceptable: 'I should not, all things considered, omit to add that in presenting this work *viva voce* [in 1855], I had proposed for the *Alca impennis* the name *Gyralca,* on the hypothesis that the name of *Pinguinus,* which I regard as better, could not be applied.'

While European writers tended to retain the name *Alca impennis* on the grounds of customary usage, for many years American writers employed the name *Plautus.* In 1917, however, it was decided that *Plautus* should be transferred to the Little Auk (Dovekie). A challenge was made to this redesignation in 1969 on the grounds that its use (by Gunnerus in 1761) had not been strictly binomial. As a result a decision was taken by the International Committee for Zoological Nomenclature to suppress *Plautus* altogether. The reasoning behind this drastic action was that if *Plautus* could not after all be applied to the Little Auk, strict application of the rules would mean that it should be applied once more to the Great Auk. Understandably, it was felt that this would be to defeat the object of having strict rules for taxonomic classification — namely, to ensure stability and consistency whenever these names were cited in books or academic papers. As it was, ever since *Plautus* had been transferred to the Little Auk following the recommendation of C.W. Richmond (*Proc. U.S. Nat. Mus.* **53**:565ff.) some writers had employed Bonnaterre's *Pinguinus* for the Great Auk, others had continued with the by then well-established *Plautus,* while others still — notably the authors of the *Handbook of British Birds* (1938–41) — had stuck doggedly to *Alca.* The position adopted by the American Museum of Natural History was succinctly summarized by E. Eisenmann in 1974 (*Auk* **91** (**2**):432):

> Wetmore and Watson (1969 *Brit. Bull. Orn. Club* **89**:6–70) pointed out that *Plautus* or *Plotus* Gunnerus 1761, then currently used as the generic name of the Dovekie (or Little Auk) was invalid, thus making available *Plautus* Brünnich 1772 as the earliest generic name for the Great Auk. To avoid a confusing transfer of names and to validate the current use of *Pinguinus* Bonnaterre 1791, an application was made to the International Committee for Zoological Nomenclature to suppress *Plautus* or *Plotus* of whatever authorship and to place on the Official List of generic names *Pinguinus* for the Great Auk and *Alle* Link 1807 for the Dovekie — the usage recommended in the 32nd Supplement to the American Ornithologists' Union *Checklist of North American Birds* 1973 (*Auk* **90**:411–419). The International Committee by Opinion 999 has approved the application (1973 *Bull. Zool. Nomen.* **30** Part 2:80–81).

Much as there is reason to be satisfied that stability in the use of a generic name for the Great Auk has at long last been achieved, we may wonder how durable *Pinguinus* will prove to be. If the Great Auk and the Razorbill truly belong to different genera then, as Steenstrup suggested in 1857, it is singularly apt, not least because the Great Auk was the first species to be termed Penguin by mariners and fishermen. At least some of those who have advocated retention of the name *Alca*, however, have done so not on account of a sentimental attachment to a name which was universally employed in the era when the fogs of ignorance surrounding the Great Auk's range and affiliations were being lifted, but because, like the comparative anatomist F.A. Lucas, they have considered the separation of the Great Auk and the Razorbill into two genera to be wholly unwarranted. Lucas concluded his *Great Auk Notes* in volume 5 of *The Auk* (1888) in characteristic vein:

> Having just compared three mounted skeletons [of the Great Auk] with one of the Razorbill, the conclusion is unavoidable that the two species resembled one another very closely in outward contour.
>
> As for internal structure, I must plead guilty to a belief that the two species should be included in the genus *Alca*, and with this bit of Cis-Atlantic heresy bring these notes to a close.

In 1977 an important paper by Storrs L. Olson (*Proc. Biol. Soc. Washington* **90** (3):690–7) gave details of the discovery at Lee Creek, North Carolina of fossil bones dating from the Pliocene epoch (2–5 million years ago) of a close relative, perhaps the immediate ancestor, of the Great Auk, which was designated by the author under the name *Pinguinus alfrednewtoni*. By adopting the name *Pinguinus* Olson necessarily strengthened the argument in favour of retaining the Great Auk in a separate genus from its smaller relative, the Razorbill. In a paper published in 1985 and entitled 'The Phylogeny of the Alcidae' (*Auk* **102**:520), however, J.G. Strauch, Jnr. observed that there were no qualitative differences in the wing structure of the two species apart from those indicating an adaptation to a different mode of feeding. Moreover, this point could be corroborated by reference to the discoveries described by Olson, for the fossil bones appear to be intermediate between the two species.

Evidently controversy will not end entirely until comparative anatomists agree what changes to bone structure brought about by adaptation to a particular mode of life are sufficient to warrant distinguishing these, or any closely related birds, not by species alone but by genus. This has remained a contentious issue for over a century.

Appendix 3

Act for the Preservation of Sea Birds

The following constitutes an abridged version of the Act which received Royal Assent on 24 June 1869.

> Whereas the sea birds in the United Kingdom have of late years greatly decreased in number, it is expedient therefore to provide for their protection during the breeding season: Be it enacted...that:
>
> *Section 2.* Any person who shall kill, wound or attempt to kill, wound or take any sea bird or use any boat, gun, net, or other engine or instrument for the purpose of killing, wounding or taking any sea bird, or shall have in his control or possession any sea bird recently killed, wounded or taken between the first day of April and the first day of August in any year, shall, on conviction of any such offence before any Justice or Justices of the Peace in England or Ireland, or before the Sheriff or any Justice or Justices of the Peace in Scotland, forfeit and pay for every such sea bird so killed, wounded or taken, or so in his possession, such sum of money not exceeding One Pound as to the said Justices or Sheriff shall seem meet, together with the costs of the conviction, provided always that this section shall not apply where the said sea bird is a young bird unable to fly.
>
> *Section 4.* Where any person shall be found offending against this Act, it shall be lawful for any person to require the person so offending to give his Christian name, surname, and place of abode; and in case the person offending shall, after being so required, refuse to give his real name or place of abode, or give an untrue name or place of abode, he shall be liable, on being convicted of any such offence...to forfeit and pay, in addition to penalties imposed by Section 2, such sum of money not exceeding Two Pounds, as to the convicting Justice or Sheriff shall seem meet, together with the costs of the conviction.
>
> *Section 5.* One moiety of every penalty and forfeiture under this Act shall go and be paid to the person who shall inform and prosecute for the same, and the other moiety shall...be paid to some one of the overseers of the poor, or to some other officer...of the parish, township or place in which the offence shall have been committed.

Section 8. The operation of this Act shall not extend to the island of St. Kilda.

Section 9. It shall be lawful for Her Majesty, by Orders in Council, where, on account of the necessities of the inhabitants of the more remote parts of the sea coasts of the United Kingdom, it shall appear desirable, from time to time, to exempt any part or parts thereof from the operation of this Act; and every such Order shall assign the limits of such part or parts aforesaid within which such exemption shall have effect.

Appendix 4

The Victorian egg collectors

In the nineteenth century there can have been few naturalists who were not also collectors of either specimens or eggs or both. What follows is a glimpse of the steps it was sometimes necessary to take when endeavouring to build up a comprehensive collection:

> One day Mr Hancock received a visit from the Edinburgh birdstuffer, Mr Small, who, on being shown a tray of [great] auks' eggs and plaster casts,[3] stated that he had recently seen a sailor offering one for £3; the sailor had been told that it was valuable, when at Liverpool: he was carrying it in his jacket pocket upon a string, like an ostrich's egg. On hearing this, the late Mr Wolley, who chanced to be in Newcastle, without loss of time started in chase, and dogged the sailor with indefatigable perseverance, coming up with him just too late, for the man was drunk, and the egg lost or broken.
>
> J.H. Gurney, *Zoologist* (1868)

Alfred Newton was the beneficiary of John Wolley's will when the latter died prematurely in 1859. Newton described taking possession of Wolley's collection in the following terms:

> There were twenty four enormous packages which weighed altogether one ton and filled a railway truck — not a single breakage. After consulting on the subject with P.L. Sclater [the editor of *The Ibis*] I came to the conclusion that I should most advantageously be serving the interest of Ornithology by publishing from Mr Wolley's notebooks a complete catalogue of the contents of his egg cabinets.

Eventually this was duly done, the *Ootheca Wolleyana* being edited in stages (Part I appeared in 1862; Part II not until 1902; Part III in 1905; and Part IV in 1907), chiefly towards the end of Newton's life. During the intervening years Newton was busy with writing or editing other encyclopaedic works on ornithology at which he worked methodically but with exasperating slowness.

As egg collecting has been illegal in the United Kingdom since 1952, and was regarded on both sides of the Atlantic as cruel and destructive even in the late nineteenth century, it is appropriate to cite a little anecdote which clearly gives an insight to Alfred Newton's attitude to the question:

> None of those who were present are likely to forget the occasion one evening in Newton's rooms [at Cambridge] when a young man interrupted an interesting talk on the fate of (it may have been) Moas with the rather large question, 'Why *do* birds become extinct?' The Professor replied without hesitation: 'Because people don't observe the Game Laws; see Deuteronomy xxii: 6.' The conversation languished after that and we soon returned to our various colleges where we looked up his reference and read: 'If a bird's nest chance to be before thee in the way of any tree or on the ground, whether they be young ones or eggs and the dam sitting upon the young or upon the eggs, thou shalt not take the dam with the young. But thou shalt in any wise let the dam go and take the young to thee that it may be well with thee that thou mayest prolong thy days.'

A.F.R. Wollaston's *Life of Alfred Newton* (1921)

Regarding Great Auk eggs in Alfred Newton's own collection, which he shared with his brother Edward, the facts are as follows: he purchased one egg from a dealer in the Strand, London; received two as part of John Wolley's legacy; and, later in life, was given four by his friend Lord Lilford, two of which, in rather poor condition, had sold in Edinburgh as late as 1880 for as little as thirty-two shillings. Unfortunately, the provenance of but one of these eggs can be traced with certainty: one of those left to him by John Wolley had been acquired from a dealer in Hamburg by John Gould, having been taken on Eldey, Iceland, in 1834 or 1835.

In addition to being regarded as the nation's foremost oologist, Alfred Newton was famous for discovering, in December 1861, no less than ten eggs of the Great Auk gathering dust in the eighteenth-century collection of John Hunter (author and once renowned Surgeon–General to the Army) which was housed at the Royal College of Surgeons, London. 'I mean to have them,' Newton had declared at once, but in this he was to be disappointed — probably for the reason given here, that other collectors were initially rebuffed by the College authorities. Nevertheless, one of the four eggs given to him later by Lord Lilford did come from that collection via Lilford's brother-in-law, who must have had greater powers of persuasion. The fact that Newton had also found at the time of his discovery a box inscribed 'Penguin eggs — Dr Dick' suggests that Hunter had acquired them in, or from, North America. The bulk of the rest of John Hunter's collection of 10 563 specimens (which had, it may be remarked, been acquired with public money by the Government in 1795) was destroyed by a bomb during the Second World War. A notable collection made by his elder brother William (1718–83) remains in the care of Glasgow University.

The following description by R. Champley of Scarborough of his own collection of Great Auk eggs, all acquired by purchase, was forwarded to

Symington Grieve, the author of *The Great Auk or Garefowl* (1885). It is of interest not only for illustrating the irresistible allure that such eggs had in the decades immediately after it became clear that the Great Auk was extinct, but also for revealing the provenance of a significant percentage of those eggs which yet survive, now, for the most part, in public collections.

No 1, figured in Thienemann

The above egg, and the first obtained, came into my possession in the following singular manner. I had written to Mr Newman, editor of *The Zoologist*, a letter asking him if he could inform me what had become of the egg that belonged to the late Mr Yarrell.[4] I received no reply; but my letter to Mr Newman, unknown to myself, was inserted on the outer cover of the *Zoologist.* Some months later I received a letter from Kunz, Leipzig. He informed me that he had seen my query respecting Yarrell's egg and said he had an egg for sale. Would I have it? He wrote a second letter giving me the price, £18 (July, 1859). Five letters refer to this transaction. The egg, from its beautiful shape, is the finest known. This egg was purchased from Theodore Schulz in 1857, he then residing in Neuhaldensleben, Saxony, a short description of which appears in *Cabanis* (Jan. 1860). Schulz purchased it from a person of the same name then residing in Leipzig. He received it, with six others, from Iceland.

No 2. This egg is engraved in Bädeker's *European Öology*

My first intimation of its whereabouts was from reading a number of *Cabanis.* I purchased the egg with the bird (said to have laid the egg) in 1861 from the apothecary Mechlenburg residing at Flinsburg, Denmark (now Germany) — the same person who sold Hancock his egg and bird obtained from Iceland, 1829.[3] Egg perfect, well marked with blotches.

No 3 — This egg was obtained during my Italian tour in 1861. I was at Verona, 31st of May, 1861. I met accidentally a Russian nobleman at the station (Porta Nueva). My acquaintance was renewed at Milan a few days later on the 2nd June. I met him near the Duomo, the day of the celebration of the unity of Italy. He told me he was going to see a certain monastery the following morning. We agreed to go together, and another friend accompanied us. We three took a carriage and pair, and arrived at the convent, fifteen miles distant, at about noon. We then drove on to Pavia, 5 miles farther. After seeing the cathedral we went to look at the university, and went over the museum of anatomy. I inquired if there were any eggs and birds in the museum and was answered in the affirmative. On looking round the glass cases, I noticed many eggs stuck on wires on shelves, but all black over with dust. I noticed among some large eggs what I thought was the egg of a great auk. I asked the attendant to open the case, but he had not the key. I told him to go for the sub-director. He returned with him and opened the case, which was fastened with screws. I took down the egg, black over with dirt, and rubbed it clean and saw it was an *Alca impennis.* I told the sub-director I would exchange some skins for it. He could not say anything but referred me to the chief director and at the same time told me that the collection was given by Prof. Spallanzani a hundred years before and that Spallanzani had been one of the lecturers at the university. My friend the Russian interpreted for me. When I obtained the address of the chief director I proceeded to his residence, accompanied by a youth, a student of the college, who spoke English — my friends meanwhile staying at the museum awaiting my return. On my arrival at the director's residence, I told him there was among the eggs an egg of 'Le Grand Pinguin' [*sic*], and I should be glad if he would let me have it for an exchange. He accompanied me back to the museum. After looking at the egg, the sub-director told him I had offered five Napoleons or an equivalent in exchange for it. They said they would rather prefer the money. I therefore borrowed the amount from my Russian friend, and, after packing the egg carefully, left the museum, they seeming sorry that they had no more specimens and considered that they had got a good bargain. We arrived at Milan at seven in the evening. I had a box made for the egg the next day. The egg is perfect and thickly pencilled at the thick end.

No 4 — This egg was obtained as follows: passing through Paris for Italy the same year (1861), I called on Parzudaki, the French naturalist. He told me the Abbé La Motte had an egg of the *Alca impennis*, but was then in Algiers. I told him to buy it for me and to write in three months to me at the Poste Restante, Rotterdam. On arriving there I found his letter, saying the son was at Abbeville and asking instructions. I at once wrote asking him to buy the egg. This he did for £24. I have four letters referring to this purchase. No history, excepting a statement that it was obtained forty years previously from French whalers.

Nos 5 & 6 — I bought these eggs in London from Ward, the naturalist in Vere St., in 1864. Previously I had received a letter from Fairmaire, Paris, saying that he had two eggs. Unfortunately his letter was sent to Scarborough while I was in London. There was consequently some delay in my knowing that he had two eggs for sale. As Fairmaire did not hear from me, he supposed I either did not care to have the eggs or that I had not got his letter. When I wrote he said he had parted with them. By chance the same week I called at Ward's, and he showed me one egg for which I gave him £25, and asked him if he had any more. He then showed me another egg for which I paid him £30. I then asked him if he had any more as I would take twenty. He smiled. He would not say how he got them, but I afterwards found out they were the same as offered to me by Fairmaire. I called on Ward many times, and he always regretted having parted with these eggs. They are perfect and well-marked. I don't know their previous history.

Nos 7, 8 & 9 — These eggs were bought in 1864 from Professor Flower, then of the Royal College of Surgeons, London. They were part of the collection of ten eggs of *Alca impennis* in the Hunterian Collection. I had difficulty in getting them as at that time they would not take money. I got over the difficulty by purchasing a collection of anatomical specimens for £45, which the museum was anxious to possess, and then exchanged it for the eggs, all very fine specimens. If I had pressed at the time, I could have got the other four eggs afterwards sold at Stevens' sale rooms in July 1865.

The following details from *The Great Auk: Records of Sales of Birds and Eggs by Public Auction in Great Britain 1806–1910* by Thomas Parkin, MA, FLS, FZS, MBOU (1911) indicates that well into the twentieth century oology was regarded as a branch of natural history which merited serious attention. This citation confirms the otherwise uncorroborated evidence given by Martin Martin in his *Late Voyage to St Kilda* (1698) regarding the colour of the eggs of the Great Auk which he stated to have been 'variously spotted, Black, Green, and Dark'.

Egg XV Auctioned June 25th 1895, Lot No 211 (slightly cracked):

Taken in Iceland about 1830 by a shipowner of St Malo who bequeathed it to Comte Raoul de Baracé whose collection was purchased by Baron d'Hamonville.

Purchased for £173 : 5 : 0d.

This is the famous egg marked with beautiful green blotches figured by Baron Louis d'Hamonville on *Pl.* 6, *Fig.* c, in the memoirs of the *Société Zoologique de France* in 1888, and of which he gives a fuller description in the Bulletin of the *Société* for 1891 (Séance de 27 janvier 1891) drawing attention to the pale green markings:

'Les taches vertes, trop peu nombreuses au gré des amateurs, qui ornent ce specimen sont très rare sur les œufs du Grand Pingouin.' (The green markings, all too seldom seen by amateurs, which adorn this specimen are very rare on the egg of the Great Auk.)

Notes

1 John Hancock was famous for his ability to make plaster casts of the eggs of the Great Auk which were all but indistinguishable from the genuine article. Many collectors, including John Wolley, engaged Hancock to make copies of the eggs in their collections.

2 William Yarrell, author of *A History of British Birds*, the standard reference work of the mid-nineteenth century. Alfred Newton's biographer, A.F.R. Wollaston (*op. cit.*) related what became of the subject of Champley's initial enquiry: 'The history of Mr Yarrell's egg which went to Mr F. Bond [a dealer] and subsequently into the the collection of Baron d'Hamonville, was investigated by Mr [Edmund] Harting. Yarrell [who died in 1856] had bought the egg as a duck's egg from a fisherwoman at Boulogne or Paris whose story was

that she had received from her husband, who had been a seaman on board a whaler, implying that it might have been brought from the Arctic regions. [Although this destination may have been *inferred* by Newton, it is likely that the French widow implied simply that the whaler on which her husband served had been operating in the waters off Newfoundland or Iceland.] This Newton considered most improbable… "It is only on a point like this which one has been driving into people for more than thirty years, that I feel called upon to interfere; but I see that the attempt is useless; though it does vex me, I confess, when those who ought to know, and really do know, better, inconsiderately help to maintain the popular delusion [that the Great Auk was found above the Arctic Circle]. This delusion was for a long while (and possibly is now) shared by Mr Champley of Scarborough and I know at one time he was busy in enquiring of Arctic navigators, and others who had been in the Far North, after *Alca impennis*, though the absurdity of such an enquiry had been demonstrated for several years."'

3 This date, if accurate, implies that the specimen came from the Geirfuglasker the year before the eruption which submerged the entire island. It seems more likely that it came, like so many others at this time, from Eldey some years after 1830. Newton, writing in the *Ibis* in 1861, fixed the date unequivocally as being 1834 when nine birds were taken there.

Appendix 5

Other bird species

An alphabetical list of the birds mentioned in the text, other than the Great Auk *Pinguinus impennis*, with their species names.

American Wigeon	*Anas americana*
Ancient Murrelet	*Synthliboramphus antiquus*
Arctic Tern	*Sterna paradisaea*
Black Guillemot	*Cepphus grylle*
Black-throated Diver	*Gavia arctica*
Black-headed Gull	*Larus ridibundus*
Brent Goose	*Branta bernicula*
Broad-billed Sandpiper	*Limicola falcinellus*
Brünnich's Guillemot	*Uria lomvia*
Bustard, *see* Great Bustard	
Buzzard	*Buteo* (sp.)
Common Guillemot	*Uria aalge*
Cormorant	*Phalacrocorax carbo*
Cornish Chough	*Pyrrhocorax pyrrhocorax*
Corncrake	*Crex crex*
Crane	*Grus grus*
Craveri's Murrelet	*Synthliboramphus craveri*
Crow (Hooded)	*Corvus corone cornix*
Cuckoo	*Cuculus canorus*
Dodo	*Raphus cucullatus*
Dovekie, *see* Little Auk	
Eagle (White-tailed)	*Haliaeetus albicilla*
Eagle (Golden)	*Aquila chrysaetos*
Eider	*Somateria mollissima*
Emperor Penguin	*Aptenodytes forsteri*
Eskimo Curlew	*Numenius borealis*

Fulmar	*Fulmarus glacialis*
Gannet	*Morus bassanus*
Gentoo Penguin	*Pygoscelis papua*
Golden Eagle	*Aquila chrysaetos*
Goose	*Anser* (sp.)
Great Bustard	*Otis tarda*
Great Northern Diver	*Gavia immer*
Great White Heron	*Egretta alba*
Guillemot *see* Common Guillemot	
Gyr Falcon	*Falco rusticolus*
Heron	*Ardea* (sp.)
Jack Snipe	*Lymnocryptes minimus*
Japanese Murrelet	*Synthliboramphus wumizusume*
Jay	*Garrulus glandarius*
Kestrel	*Falco tinnunculus*
Kittiwake	*Rissa tridactyla*
Labrador Duck	*Camptorhynchus labradorius*
Leach's Storm-petrel	*Oceanodroma leucorhoa*
Little Auk	*Alle alle*
Magpie	*Pica pica*
Mallard	*Anas platyrhynchos*
Merganser (Red-breasted)	*Mergus serrator*
Murre	*Uria* (sp.)
Oystercatcher	*Haematopus ostralegus*
Passenger Pigeon	*Ectopictes migratorius*
Peregrine Falcon	*Falco peregrinus*
Petrel	*Procellarius* (sp.)
Pheasant	*Phasianus colchicus*
Pine Grosbeak	*Pinicola enucleator*
Plover (European Golden)	*Pluvialis apricaria*
Puffin	*Fratercula arctica*
Raven	*Corvus corax*
Razorbill	*Alca torda*
Red Grouse	*Lagopus lagopus*
Red-throated Diver	*Gavia stellata*
Red-throated Pipit	*Anthus cervinus*
Ringed Plover	*Charadrius hiaticula*
Roseate Tern	*Sterna dougallii*

Sandwich Tern	*Sterna sandwicensis*
Shag	*Phalacrocorax aristotelis*
Shearwater	*Puffinus* (sp.)
Shorelark	*Eremophila alpestris*
Skua(s)	*Stercorcariidae*
Smew	*Mergus albellus*
Solan Goose, *see* Gannet	
Spotted Redshank	*Tringa erythropus*
Storm Petrel	*Hydrobates pelagicus*
Teal	*Anas crecca*
Temminck's Stint	*Calidris temminckii*
Thrush	*Turdus* (sp.)
Waxwing	*Bombycilla garrulus*
Wheatear	*Oenanthe oenanthe*
White Owl	*Tyto alba*
Wren	*Troglodytes t. hirtensis*
Whooper Swan	*Cygnus cygnus*
Wigeon	*Anas penelope*
Wild Duck, *see* Mallard	
Xantus' Murrelet	*Synthliboramphus hypoleucos*

Bibliography

Allen, J. 1876. On the extinction of the Great Auk at the Funk Is. *American Naturalist* p. 48.

Anon. 1613. Willoughby Papers, Middleton MSS (Mi X), Nottingham University Library. (Transcription by R. Barakat in *Dictionary of Newfoundland English* (ed. G.M. Story *et al.*) 1982. Toronto.)

Anon. 1762. *The American Gazetteer*. London.

Anspach, L-A. 1819. *A History of the Island of Newfoundland*. London.

Audubon, J.J.L. 1838. *Ornithological Biography* Vol. 4. Edinburgh.

Audubon, J.J.L. 1840. *The Birds of America*. New York.

Audubon, M.R. 1898. *Audubon and his Journals*. New York.

Audubon, J.J.L. 1930. *Letters of John James Audubon* (ed. H. Corning). Boston.

Baikie, W.B. and Heddle, R. 1848. *Historia Naturalis Orcadensis*. Edinburgh.

Banks, Sir Joseph 1971. *Joseph Banks in Newfoundland and Labrador, 1766* (ed. A.M. Lysaght). London.

Baring-Gould, S. 1863. *Iceland: its Scenes and Sagas*. London.

Benicken, F. 1824. Beiträge zur nordischen Ornithologie. *Isis* 877–91. Jena.

Birkhead, T.R. 1993. *Great Auk Islands*. London.

Birkhead, T.R. 1994. How collectors killed the Great Auk. *New Scientist* 28th May. London.

Blasius, W. 1884. Zur Geschichte der Ueberreste von *Alca impennis*. *Journal für Ornithologie* pp. 58–176.

Bones, M. 1993. The Garefowl or Great Auk. *Hebridean Naturalist* **11**:15–24.

Bonnycastle, Sir Richard 1842. *Newfoundland in 1842*. London.

Bourne, W.R.P. 1993. The story of the Great Auk. *Arch. Nat. Hist.* **20** (2):257–78.

Brisson, M.J. 1760. *Ornithologia sive Synopsis Methodica Avium* Vol. 6. Paris.

Brown, J.A.H. and Buckley, T.E. 1891. *Vertebrate Fauna of Scotland* Vol. 4. Edinburgh.

Buffon, Comte de (Leclerc, G.L.) 1770. *Histoire naturelle des oiseaux* Vol. 9. Paris.

'C., T.' 1623. *A Short Discourse of the New Found Land.* Dublin.

Cartwright, G. 1792. *Journal of Transactions and Events…on the Coast of Labrador* [*etc.*]. Newark.

Catesby, M. 1731–43 (1748). *The Natural History of Carolina, Florida and the Bahama Islands.* London.

Charlton, E. 1860. On the Great Auk. *Zoologist* (1st series) pp. 6883–8. London.

Clarke, W.E. 1912. *Studies in Bird Migration.* London and Edinburgh.

Clusius, C. (L'Écluse, C. de) 1605. *Exoticorum Libri Decem* Book 5. Leyden.

Coish, E.C. 1994. *Distant Shores.* St John's.

Coues, E. 1868. Monograph of the Alcidae. *Proc. Acad. Nat. Sci. Philadelphia* p. 20.

Coward, T.A. 1919. *The Birds of the British Isles and their Eggs.* London and New York.

Coward, T.A. 1931. *Bird and Other Nature Problems.* London.

Deacon, R. 1966. *Madoc and the Discovery of America.* London.

Debes, L.J. 1676. *Feroœ & Feroa Reserata* [*etc.*] (trans. J. Sterpin). London.

Denys, N. 1672. *The Description and Natural History of the Coast of North America (Acadia)* (trans. W.F. Ganong. 1908). Toronto.

Dixon, C.E. 1894. *The Nests and Eggs of Non-Indigenous British Birds.* London.

Dunning, J.W. 1872. Great Auk. *Zoologist* (2nd series) p. 2946. London.

D'Urban, W. and Mathew, M. 1892. *The Birds of Devon.* London.

Edwards, G. 1750. *The Natural History of Uncommon Birds* Vol. 3. London.

Faber, F. 1822. *Prodromus der Isländischen Ornithologie.* Copenhagen.

Faber, F. 1826. *Ueber das Leben der hochnordischen Vögel.* Leipzig.

Faber, F. 1827. Seereise nach den Vögelscheeren. *Isis* 685–8. Jena.

Fabricius, O. 1780. *Fauna Groenlandica.* Hafniae et Lipsiae.

Feilden, H.W. 1872. The Birds of the Faeroe Islands. *Zoologist* (2nd series) pp. 3280–5. London.

Fisher, J. 1947–51. *Bird Recognition.* London.

Fleming, J. 1824. Gleanings of natural history during a voyage along the coast of Scotland in 1821. *Edinb. Phil. Journ.* **10**:95–101.

Fleming, J. 1828. *History of British Mammals.* Edinburgh.

Forster, J.R. 1786. *History of the Voyages and Discoveries made in the North.* London.

Foss, M. 1974. *Undreamed Shores: England's wasted Empire in America.* London.

Fuller, E. 1999. *The Great Auk.* Published by the author, Southborough, Kent.

Gardner, Dr (of Boston) *c.*1790. *Newfoundland.* Unpublished MS, British Library No. 15 493.

Garrad, L.S. 1972. Bird remains including those of a Great Auk…at Perwick Bay. *Ibis* **114**:258–9. B.O.U.

Garrad, L.S. 1972. *The Naturalist in the Isle of Man.* Newton Abbot.

Gaskell, J.M. 1999. A review of some early testimony from the New World in respect of the Great Auk (etc.). *Arch. Nat. Hist.* **26** (**1**):101–12.

Gordon, S. 1962. *Highland Days.* London.

Gould, J. 1837. *The Birds of Europe* Vol. 5. London.

Graba, C-J. 1830. *Tagebuch, geführt auf einer Reise nach Färö im Jahre 1828.* Hamburg and Kiel.

Gray, R. 1871. *Birds of the West of Scotland.* Edinburgh.

Gray, R. 1880. On two unrecorded eggs of the Great Auk [etc.] *Proc. Roy. Soc. Edinb.* **10**:668–82.

Grieve, S. 1885. *The Great Auk or Garefowl.* Edinburgh.

Grieve, S. 1888. Recent notes on the Great Auk. *Edinb. Field Nat. and Microscop.* Soc. **2**:92–119.

Gurney, J.H. 1868 The Great Auk. *Zoologist* (2nd series) pp. 1442–53. London.

Hakluyt, R. 1600. *The Principall Navigations…of the English Nation* [*etc.*] Vol. 3. London.

Hakluyt Society. 1940. *The Voyages and Colonising Enterprises of Sir Humphrey Gilbert.* London.

Hansard (Publishers). 1868–9. *Parliamentary Reports.* London.

Harris, M.P. and Birkhead, T.R. 1985. Breeding ecology of the Atlantic Alcidae in *The Atlantic Alcidae* (ed. D.N. Nettleship and T.R Birkhead). London and New York.

Harvey, M. 1874. The Great Auk. *Forest and Stream* **2:** No 16:386. New York.

Henderson, E. 1819. *Iceland, or the Journal of a Residence in that Island 1814–15.* Edinburgh.

Herman, O. 1905. *The Method of Ornithophaenology.* Budapest.

Hewitson, W. 1853–6. *Coloured Illustrations of the Eggs of British Birds.* London.

Hobson, K.A. and Montevecchi, W.A. 1991. Stable isotopic determinations of trophic relationships of Great Auks. *Oecologia* **87**:528–31.

Horrebov, N. 1758. *The Natural History of Iceland.* London.

Jardine, Sir W. (ed.) 1848–53. *Contributions to Ornithology.* London.

Josselyne, J. 1672. *New England's Rarities Discovered* [*etc.*]. London.

Kingsley, C. 1863. *The Water Babies.* London.

Kjærbölling, N. 1851–6. *Ornithologia Danica.* Copenhagen.

Köppen, W. 1900. Versuch einer Klassification der Klimata, vorzugsweise nach ihren Beziehung zur Planzenwelt. *Geog. Zeitschr.* 593–611.

Landt, J. 1800. *Forsøg til en Beskrivelse over Faeroerne.* Copenhagen.

Landt, J. 1810. *A Description of the Faeroe Islands* [*etc.*] (translated from the Danish). London.

Latham, J. 1790. *Index Ornithologicus* [*etc.*]. London.

Lescarbot, M. 1609. *Histoire de la nouvelle France* [*etc.*]. Paris.

Lloyd, L. 1854. The Great Auk still found in Iceland. *Edinb. New. Philos. Journ.* **56**:260–2.

Loyd, L.R.W. 1925. *The History and Natural History of Lundy.* London.

Lucas, F.A. 1888. Great Auk notes. *Auk* **5**:278–83.

Lucas, F.A. 1890. The Expedition to Funk Island. *Rep. U.S. Nat. Mus.* 1887–8.

Lucas, F.A. 1891. Explorations in Newfoundland and Labrador in 1887. *Rep. U.S. Nat. Mus.* 1888–9.

Lucas, F.A. 1901. *Animals of the Past.* London.

Macaulay, K. 1764. *The Story of St. Kilda.* London.

Macaulay, K. 1765. *A Voyage to and History of St. Kilda.* Dublin.

Macgillivray, W. 1852. *A History of British Birds* Vol. 4. London.

McClintock, Sir F.J. 1860. The Great Auk. *Zoologist* (1st series) p. 6981. London.

Martin, M. 1698. *A Late Voyage to St. Kilda.* London.

Martin, M. 1934. *A Description of the Western Islands of Scotland* [*etc.*] (ed. D.J. Macleod). Stirling.

Mason, J. 1620. *A Briefe Discourse of the New-found-land.* Edinburgh.

Mathew, M.A. 1866. The Great Auk on Lundy Island. *Zoologist* (2nd series) pp. 100–1.

Meldgaard, M. 1988. The Great Auk...in Greenland. *Hist. Biol.* **1**:145–78.

Micahelles, K. 1833. Zur Geschichte der *Alca impennis. Isis* 648–51. Jena.

Milne, J. 1875. Relics of the Great Auk on Funk Island reprinted from *The Field.* London.

Mitchell, W. 1971. *A Few Million Birds.* London.

Mohr, N. 1786. *Forsœg til en Islandsk Naturhistorie [etc.].* Copenhagen.

Munro, Sir Donald 1549. Description of the Western Islands of Scotland in Montagu, G. 1802 and 1813. *Ornithological Dictionary* (and *Supplement*). London.

Montevecchi, W.A. and Tuck, L.M. 1987. *Newfoundland Birds* [*etc.*]. Cambridge, Mass.

Morris, F.O. 1868 (Aug. 29th). Letter to *The Times.* London.

Nelson, T.H. 1907. *The Birds of Yorkshire* Vol. 2. London.

Nettleship, D.N. and Evans, P.G.H. 1985. Distribution and status of the Atlantic Alcidae in *The Atlantic Alcidae* (ed. D.N. Nettleship and T.R. Birkhead). London and New York.

Newton, A. 1860. Memoir of John Wolley. *Ibis* (2). B.O.U.

Newton, A. 1861. Abstract of Mr J. Wolley's researches in Iceland [etc.]. *Ibis* (3). B.O.U.

Newton, A. 1862. *The Zoology of Ancient Europe.* London and Cambridge.

Newton, A. 1862. Remarks on the exhibition of a natural mummy of the Great Auk. *Proc. Zool. Soc. London* 435–8.

Newton, A. 1865. The Garefowl and its Historians. *Nat. Hist. Review* 467–78. London.

Newton, A. 1864–1907. *Ootheca Wolleyana.* London.

Ogilby, J. 1671. *America.* London.

Olafsson, E. and Paulson, B. 1802. *Voyage en Island.* Paris and Strasbourg.

Olson, S.L. 1977. A Great Auk (*Pinguinus*) from the Pliocene of North Carolina. *Proc. Bio. Soc. Washington* **90**:690–7.

Olson, S.L., Swift, C.C., and Mokhiber, C. 1979. An attempt to determine the prey of the Great Auk. *Auk* **96**:790–2.

Orton, J. 1870. The Great Auk. *American Naturalist* 539–42.

Osborn, F. 1948. *Our Plundered Planet.* London.

Ossa, H. 1973. *They Saved Our Birds.* New York.

Owen, R. 1865. Description of the skeleton of the Great Auk or Garefowl. *Trans. Zool. Soc. London* **5**:317–35.

Parkin, T. 1911. The Great Auk — a record of sales (etc.) *Hastings & E. Sussex Naturalist* **1** (extra paper).

Peckham, Sir George 1583. *A True Reporte of the late discoveries…of the New-found Landes [etc.].* London.

Pedley, C. 1863. *History of Newfoundland.* London.

Pennant, T. 1784. *Arctic Zoology* Vol. 2. London.

Perry, R. 1940. *Lundy Isle of Puffins.* London.

Perry, R. 1978. *Wildlife in Britain and Ireland.* London.

Piatt, J.F. and Nettleship, D.N. 1985. Diving depths of four Alcids. *Auk* **102**:293–7.

Pinkerton, J. 1808–14. *General Collection of Voyages.* London.

Prowse, D.W. 1895. *History of Newfoundland.* London.

Raven, C.E. 1925 (1950). *In Praise of Birds.* London.

Ray, J. 1678. *The Ornithology of Francis Willughby.* London.

Reeks, H. 1869. Notes on the zoology of Newfoundland. *Zoologist* (2nd series) pp. 1854–6. London.

Reeves, J. 1793. *History of the Government of Newfoundland.* London.

Rennie, J. 1835. *The Faculties of Birds.* London.

Sage, B. *et al.* 1986. *The Arctic and its Wildlife.* London.

Sagard-Théodat, G. 1632. *Le Grand Voyage au pays des Hurons … dite Canada.* Paris.

Saunders, H. (ed.) 1871. Yarrell's *A History of British Birds* (2nd edn) Vols. 3–4. London.

Selby, P.J. 1825–32. *Illustrations of British Ornithology.* Edinburgh.

Sibbald, Sir Robert 1683. *Scotia Illustrata.* Edinburgh.

Sigurdsson, G. (attrib.) *c.*1710. Geirfuglasker MS Lbs. 44, Fol. Bl. 71–6. National Library of Iceland.

Skene, W. 1876–80. *Celtic Scotland.* Edinburgh.

Smith, J.A. 1878–80. Notice on the remains of the Great Auk…with notes on its occurrence in Scotland; additional notes. *Proc. Soc. Antiq. Scot.* Vols. 13–14.

Steenstrup, J.S. 1857. *Et Bidrag til Geirfuglens…Naturhistorie* (*etc.*). Copenhagen. (French translation: *Bull. Soc. Orn. Suisse* (1868) **2**:5–70.)

Strauch, J.G., Jr. 1985. The Phylogeny of the Alcidae. *Auk* **102**:520–39.

Temminck. C.J. 1820–40. *Manuel d'Ornithologie* (2nd edn). Paris.

Thevet, A. 1558. *Les singularitez de la France antarctique.* Paris.

Thevet, A. 1568. *The New found Worlde, or Antarctike…travailed and written in the French tong by … Andrewe Thevet.* (trans. T. Hacket) London.

Thomas, A. 1795. *The Newfoundland Journal of Aaron Thomas* (ed. J. Murray (1968)). London.

Thompson, W. 1851. *The Natural History of Ireland* Vol. 3. Dublin.

Tocque, P. 1878. *Newfoundland as it was and as it is in 1877*. London and Toronto.

Townsend, C.W. (ed.) 1911. *Captain Cartwright and his Journal*. London and Boston.

Tristram, H.B. 1860. *Address to the Tyneside Field Naturalists' Club*. Newcastle.

Tunstall, M. 1771. *Ornithologia Britannica*. London.

Tuck, L.M. 1960. *The Murres*. Canadian Wildlife Monograph Series, Ottawa.

United Society for the Propagation of the Gospel. *Calendar of Letters 1721–93*.

Violani, C. 1974. Ecologia di un'estinzione: l'*Alca impenne*. *Boll. Mus. Civ. Stor. Nat. Venezia* **25**: 49–60.

Wallis, J. 1769. *The Natural History and Antiquities of Northumberland*. London.

Whitbourne, R. 1620. *Discourse and Discovery of Newfoundland*. London.

Whitbourne, R. 1622. *Discourse Containing a Loving Invitation [etc.]*. London.

Williams, G.A. 1979. *Madox, the making of a myth*. London.

Williamson, K. 1939. A Manx record of the extinct Great Auk. *Journ. Manx Mus.* **4**:168–72.

Williamson, K. 1948. *Atlantic Islands*. London.

Willughby, F. 1676. *Ornithologia Libri Tres*. London.

Witherby, H.A. *et al.* 1938–41. *The Handbook of British Birds*. London.

Wix, E. 1836. *Six Months of a Newfoundland Missionary's Journal*. London.

Wollaston, A. 1921. *Life of Alfred Newton*. London.

Wolley, J. 1850. Some observations on the birds of the Faeroe Islands in *Contributions to Ornithology* (ed. Sir W. Jardine) 1848–53. London.

Wolley, J. 1852. On the Ringed Guillemot (letter). *Zoologist* (1st series) p. 3477. London.

Wolley, J. 1858. *Garefowl Book* Vol. 1. Unpublished MS. Department of Zoology, Cambridge University.

Wood, J.G. 1862. *The Illustrated Natural History*. London.

Worm, O. 1655. *Museum Wormianum seu Historiae Rerum Rariorum*. Amsterdam.

Worth, C.B. 1940. Egg volumes and incubation periods. *Auk* **57**:44–57.

Yarrell, W. 1843. *A History of British Birds*. London.

Yonge, J. 1963. *Journal of James Yonge (1647–1721)* (ed. F.N.L. Poynter). London.

Index

Numbers in italics denote references to illustrations.